Dynamics of Biochemical Systems

NATO ASI Series

Advanced Science Institutes Series

A series presenting the results of activities sponsored by the NATO Science Committee, which aims at the dissemination of advanced scientific and technological knowledge, with a view to strengthening links between scientific communities.

The series is published by an international board of publishers in conjunction with the NATO Scientific Affairs Division

A	**Life Sciences**	Plenum Publishing Corporation
B	**Physics**	New York and London
C	**Mathematical and Physical Sciences**	D. Reidel Publishing Company Dordrecht, Boston, and Lancaster
D	**Behavioral and Social Sciences**	Martinus Nijhoff Publishers
E	**Engineering and Materials Sciences**	The Hague, Boston, and Lancaster
F	**Computer and Systems Sciences**	Springer-Verlag
G	**Ecological Sciences**	Berlin, Heidelberg, New York, and Tokyo

Recent Volumes in this Series

Series A: Life Sciences

Dynamics of Biochemical Systems

Edited by

Jacques Ricard

Biochemical Center, CNRS
Marseille, France

and

Athel Cornish-Bowden

The University of Birmingham
Birmingham, United Kingdom

Springer Science+Business Media, LLC

Proceedings of a NATO Advanced Research Workshop on
Dynamics of Biochemical Systems,
held September 19–23, 1983,
in Marseille, France

Library of Congress Cataloging in Publication Data

NATO Advanced Research Workshop on Dynamics of Biochemical Systems
(1983: Marseille, France)
Dynamics of biochemical systems.

(NATO ASI series. Series A, Life sciences; v. 81)
"Proceedings of a NATO Advanced Research Workshop on Dynamics of Bio-
chemical Systems, held September 19–23, 1983, in Marseille, France"—T.p.
verso.
"Published in cooperation with NATO Scientific Affairs Division."
Includes bibliographical references and index.
1. Enzymes—Congresses. 2. Biological chemistry—Congresses. I. Ricard,
Jacques, 1929–   II. Cornish-Bowden, Athel. III. Title. IV. Series.
QP601.N29   1983   574.19′25   84-17857
ISBN 978-1-4757-5036-2   ISBN 978-1-4757-5034-8 (eBook)
DOI 10.1007/978-1-4757-5034-8

© 1984 Springer Science+Business Media New York
Originally published by Plenum Press, New York in 1984

PREFACE

This book collects together a series of communications given
at a NATO Advanced Research Workshop held in Marseilles, France,
from September 19 to September 23, 1983. Its aim is to describe the
principles that govern the dynamics of biochemical systems. An obvious question that arises at this point is the need to define what
a biochemical system is : in this book it is defined operationally,
as a set of co-ordinated chemical reactions conditioned either by a
single enzyme existing in different conformations, or by different
enzymes, or by enzymes in association with membranes. The book thus
describes how to tackle the dynamics of biochemical systems with different levels of complexity. The simplest level of complexity is represented by different conformational states of the same enzyme that
can interact to control an enzyme reaction. An already more complex
level arises if the polypeptide chains bearing the active site are
packed together as an oligomeric enzyme. A third degree of complexity
occurs if different enzymes are packed together as a multi-enzyme complex, with added potential for channelling effects. Many enzymes in
the living cell are associated with membranes or cell envelopes, and
one may wonder how such association may modify an enzyme's behavior.
Finally, the enzyme reactions that occur in the living cell are interconnected, and constitute a very complex integrated network that
represents the highest degree of complexity.

The methods, both theoretical and experimental, for studying
biochemical dynamics originate from the kinetic study of isolated
enzymes in solution. The unifying concept in the study of biochemical systems of all levels of complexity is a physico-chemical and
quantitative study of their temporal evolution. Indeed, the concept
of evolution occupies a central position in biology, and any kind
of biological problem has to be viewed with an evolutionary perspective if one is to make sense of it. The properties of enzymes and
enzyme systems today are the result of long evolution. The last
part of the book is therefore devoted to understanding how fine tu-
ning of the behavior of enzymes and enzyme systems may have occurred
in the course of evolution.

In line with these ideas, the book comprises six sections, of which the first is entitled *"Slow" conformation changes of enzymes and their relevance to the regulation of biochemical systems*. Its aim is to show that "slow" conformation changes of an enzyme occurring far from pseudo-equilibrium conditions may generate apparent kinetic co-operativity. This behavior may occur even with monomeric one-site enzymes, and implies interaction of different conformational states of the enzyme in the overall enzyme reaction. The concepts of enzyme hysteresis and enzyme memory are related to this view. Two papers, one by K.E. Neet, G.V. Ohning and N.R. Woodruff (Cleveland, Ohio), and the other by A. Cornish-Bowden, M. Gregoriou and D. Pollard-Knight (Birmingham, England) give strong experimental support to these ideas.

The second section is entitled *Dynamics of subunit interactions in multimeric enzymes and the control of catalysis*. It describes how identical subunits interact to modulate the rate of product appearance. E.P. Whitehead (Rome, Italy) presents a linkage approach to the analysis of steady-state rates, whereas K. Dalziel (Oxford, England) offers a pre-steady-state kinetic study of polymeric oxidative decarboxylases.

The third section is devoted to the analysis of a more complex situation and is entitled *Enzyme interactions and dynamics of enzyme complexes*. Enzyme aggregation and compartmentation represent a functional advantage and may improve the performance of enzymes in the intracellular milieu. Three communications illustrate these views. G.R. Welch (New-Orleans, Louisiana) presents a rather general picture of enzyme dynamics in organized states. T. Keleti (Budapest, Hungary) describes how channelling may occur within a multi-enzyme complex, and A.H. Lane, C.H. Paul and K. Kirschner (Basel, Switzerland) present a detailed kinetic study of the conformation changes that occur in the assembly of the $\alpha_2\beta_2$ complex of tryptophan synthase.

A step further in the complexity of dynamic biochemical systems is taken in the fourth section devoted to the *Dynamics of enzyme reactions in heterogeneous media*. Under non-equilibrium conditions, diffusion of substrates and products as well as repulsion of these ligands by the fixed charges of a membrane, coupled to an enzyme reaction, may generate surprising effects. Among these are the recognition of "signals" from the external milieu (short-term memory) and the conduction of these "signals" at the surface of the membrane. Two contributions illustrate these ideas, one by J. Ricard, G. Noat and M. Crasnier (Marseilles, France) and the other by J.F. Hervagault, J. Breton, J.P. Kernevez, J. Rajani and D. Thomas (Compiègne France).

The last step in the complexity of biochemical systems is offered by the analysis of metabolic systems. This matter is aptly discussed in the fifth section of the book entitled *Dynamics of metabolic pathways: Self-organization and chaotic behavior*. The dynamic behavior of a metabolic pathway considered as a whole is quite different from that of any enzyme acting as an element of the same pathway. Small random perturbations of an external parameter may create temporal organization of the whole system. The kinds of temporal organization possible may include sustained oscillations, birhythmicity and chaos as well. Three contributions illustrate the development in this area. The first is by A. Goldbeter, J.L. Martiel and O. Decroly (Brussels, Belgium) and discusses the rhythmicity displayed by metabolic processes; the second, by B. Hess, D. Kuschmitz and M. Markus (Dortmund, Federal Republic of Germany), is specifically dedicated to the study of glycolysis; and the last , by J. Stucki (Zurich, Switzerland) discusses biological energy conversion from the point of view of non-equilibrium thermodynamics.

In the last section of the book these ideas are placed in an evolutionary context. This section, entitled *Evolutionary considerations*, is specifically concerned with three important problems : the evolution and mutation of the genetic code; the progressive transformation of binding proteins into enzymes; the kinetics of complex self-replicating molecular systems and the role played by mutation and selection in this process. The first problem is examined by J.T. Wong (Toronto, Canada), the second by B. Gutte (Zurich, Switzerland) and the last by P. Schuster (Vienna , Austria).

The content of the book is obviously interdisciplinary in character and describes research carried out at the borderline between theory and experiment. We hope that these contributions shed some light on the physical bases of the complex dynamics of biological processes.

We are glad to thank Paule Cassa, Marie-Thérèse Nicolas and Jacques Victor for their help in the preparation of the manuscript. We are especially grateful to Brigitte Videau who typed most of the contributions in camera-ready form and took a major part in the practical organization of the meeting.

Jacques RICARD

Athel CORNISH-BOWDEN

Acknowledgements

The Editors gratefully acknowledge the generous financial support furnished by NATO.

CONTENTS

SECTION I - "SLOW" CONFORMATION CHANGES OF ENZYMES AND THEIR RELEVANCE TO THE REGULATION OF BIOCHEMICAL SYSTEMS

HYSTERETIC ENZYMES, SLOW INHIBITION, SLOW ACTIVATION, AND SLOW MEMBRANE BINDING

Kenneth E. Neet, Gordon V. Ohning
and Nathaniel R. Woodruff

Case Western Reserve University, Cleveland, OH 44106
USA

INTRODUCTION

Enzymes are generally considered to utilize rapid responses since they , indeed, catalyze reactions. In fact, many early and current approaches attempt to study the rapid catalytic steps by decreasing the rate of reactions, e.g. cryoenzymology, or by increasing the instrumental methods, e.g. rapid kinetic techniques. However, it has become apparent over the past fifteen years that slower responses in enzymes may be beneficial for catalysis or regulation of the activity. Numerous enzymes have been described that undergo relatively slow changes that are manifested in non-linear progress curves, cooperativity in their steady state kinetics, or otherwise demonstrable slow structural changes in response to ligands. Early suggestions by Rabin (1967), Keleti (1967), and Witzel (1968) of the observation of slow changes during assay resulted in 1970 in the coining of the term 'hysteresis' by Frieden (1970) to describe this class of enzymes. Frieden (1970) suggested that a hysteretic enzyme with a response time of minutes might serve to dampen or buffer cellular responses to changing metabolite concentrations. Shortly threafter, we (Ainslie, et al, 1972) provided the theoretical basis that described the potential and the limitations of Ligand Induced Slow Transitions in enzymes that may contribute to the kinetic cooperativity of the steady state. Whitehead (1970) suggested that this kinetic cooperativity is through 'time' rather than 'space' as in site-site interactions. The notion that this property represented a memory of the enzyme for a previous conformation was discussed by Ricard et al, (1974) and extensively analyzed in terms of the 'mnemonic' enzyme to

3

emphasize this point. These several related concepts represent
different viewpoints of the same basic process rather than any real
conflict (Neet and Ainslie, 1980; Frieden, 1979). Several other
contributions to our understanding or application of these ideas
have been made by several investigators (Jarabak and Westley, 1974;
Whitehead , 1976; Kurganov, 1977; Storer and Cornish-Bowden, 1977).

As discussed in the published literature (Frieden, 1979; Neet
and Ainslie, 1980), the following definitions will be used. Hysteresis or hysteretic enzyme applies to any observable, slow process
affecting enzymatic activity, particularly with respect to its
potential for physiological function. Transient refers directly to
the observation of the early, nonlinear stages of an assay progress
curve, before the true steady state is attained. Slow transition
refers to the molecular change that the enzyme undergoes in its
hysteretic response, which may be a conformational change, an association-dissociation, or a ligand displacement (Frieden, 1970;
Ainslie, et al, 1972). We are concerned with reversible, non-
covalent interactions, thus eliminating the area of enzymology dealing with regulation of activity by covalent modification, e.g.
phosphorylation. 'Slow' also needs to be defined since it is a term
relative to the scale of the observer. 'Slow' may simply mean
'in the range of minutes' and thereby allow easy observations of
a transient in standard laboratory assays and potentially influence
rapid metabolic changes through its hysteretic response. Alternatively, 'slow' may refer to particular molecular steps and have certain relationships among different steps in a mechanism; e.g. , for
the Ligand Induced Slow Transition mechanism to generate cooperativity, the rate of the isomerization step must be on the same order
of magnitude as the other unimolecular steps (dissociation, catalysis) in the mechanism (Ainslie, et al, 1972).

The general form of the slow transition (isomerization) is
given (Fig. 1) for the two substrate, ordered mechanism for a monomeric enzyme in which substrate is capable of binding to two forms
of an enzyme that are slowly interconvertible. For a single substrate or conditions of saturating levels of one substrate, the
mechanism would simplify to one involving only one side of the
figure. Strictly concerted mechanisms, in which the second confor-
mer can only occur with both substrates (or products) bound, and
rapid equilibrium mechanisms, in which the binding of substrates
is at equilibrium (Ainslie, et al, 1972), can only give rise to
hysteretic transients and not to kinetic cooperativity in the
steady state. The mnemonic mechanism (Fig. 2), in which there is
one EA form and two free enzyme forms (no E'A), is shown for
comparison and is extensively discussed for a particular enzyme ,
glucokinase, in the next paper. This simpler mechanism can account
for kinetic cooperativity in a monomeric enzyme and produce transients. The observable assay, transient, and its physiological

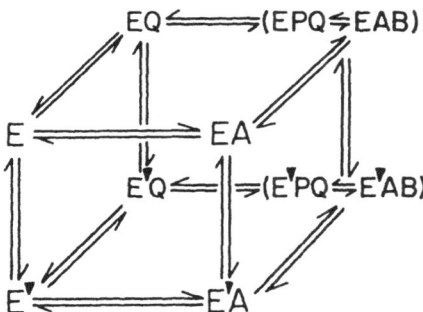

Fig.1. The Ligand Induced Slow Transition Mechanism for a hysteretic, monomeric enzyme catalyzing a two substrate, ordered reaction. The slow transition shown is an isomerization (vertical steps) between different enzyme conformations, E and E', with different catalytic properties. A and B are substrates and P and Q are products. Vertical steps are slow steps. (Reprinted by permission from Neet and Ainslie, 1980).

equivalent, is seen (Fig. 1) to be due to a time dependent change in distribution between the primed and unprimed cycles altered by the presence or change in concentration of the substrate. Cooperativity, if it exists in the monomeric enzyme, is dependent upon both a slow step as well as a shift in distribution between cycles in the steady state in a non-Michaelis-Menten fashion. Cooperativity can be positive or negative with apparent bursts or lags (or, indeed, no easily observed transient) in the transient; conditions for generating these have been discussed (Neet and Ainslie, 1980). These mechanisms have been drawn for a conformational change in a monomeric enzyme but could also be due to a slow association/dissociation (Frieden, 1970; Ainslie et al, 1972; Klinov and Kurganov, 1982), to a slow ligand dissociation (Frieden, 1970; Ainslie et al, 1972; Williams and Morrison, 1979), or occur

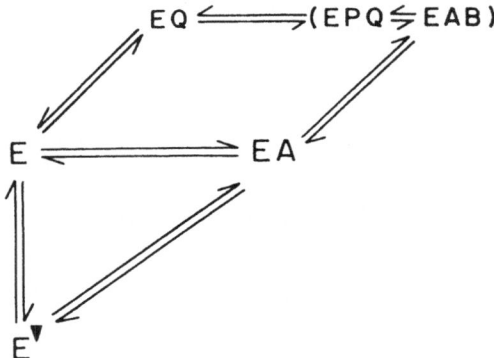

Fig. 2. The Mnemonic Mechanism for a hysteretic, monomeric enzyme
catalyzing a two substrate reaction.E and E' are two different enzyme conformations that convert to the same catalytic form upon
binding substrate, A. An ordered reaction is shown. A,B, P and Q
have the same meaning as Fig. 1.

in an oligomeric enzyme that might also display cooperativity due
to site-site interactions (Kurganov, 1982). Sorting out the contribution, significance, and interaction of the hysteretic response
in many enzymesis the subject of important, current investigations.

CONCEPTS RELATED TO HYSTERESIS

Slow inhibition

Slow-binding inhibitors are those in which the establishment
of equilibrium between E, I and EI complexes does not occur instantaneously but occurs in the time range of minutes (Cha, 1975;
Williams and Morrison, 1979). Such properties were implicit in the
observation and formulation of early hysteretic and slow transition concepts (Frieden, 1970; Ainslie, et al, 1972). The observable , slow step could either be the initial encounter complex
formation or a slow transition to an altered EI' complex after

a rapid initial binding step. In the latter case the overall
equilibrium constant would be the resultant of both steps and could
lead to a quite high affinity. If the affinity is high enough, then
significant depletion of the free inhibitor concentration could
also result and the effector would then be operationally classified
as a tight binding inhibitor (Williams and Morrison, 1979; Morrison
1982), i.e. one with which stoichiometric ratios of inhibitor and
enzyme are used during experimental analysis. A slow binding inhibitor can be one type of tight binding inhibitor if its affinity
is high enough (or the conditions of analysis are appropriate) or
it may be a relatively weak inhibitor but with a slow onset of the
full inhibitory state. Progress transients may be 'bursts' or 'lags'
depending upon the preincubation and assay conditions. Analysis of
the non-linear transient curves and consideration of possible arti-
facts (Williams and Morrison, 1979; Morrison, 1982) are similar to
those discussed for substrate-induced, hysteretic enzymes
(Frieden, 1979; Neet and Ainslie, 1980).

    Mechanistically, the simple, competitive case or dead-end
inhibition (Fig.3A) is the most straightforward and can be directly

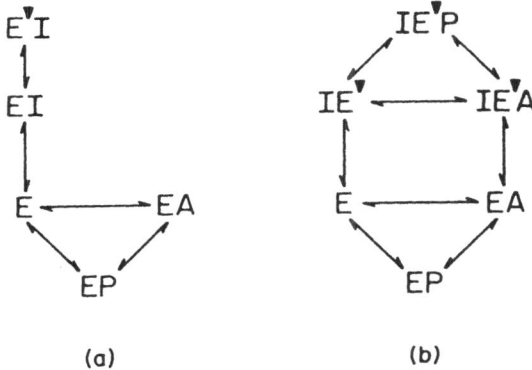

(a)   (b)

Fig. 3. (a) A slow, competitive inhibition mechanism for the conversion of A to P by the enzyme E. The inhibitor (I) binds only to the
free enzyme form. (b) A more general slow inhibition scheme which
can give rise to non-competitive or mixed inhibition. The binding
of inhibitor, I, and the putative subsequent isomerization are
shown as one step for simplicity.

analyzed for the rate constants in the pre-steady state (Cha, 1975).
The more complex situation in which the EIA' complex is catalytically competent (Fig.3B)can lead to a non-competitive or mixed inhibition and a more complex relationship between the relaxation time
for the transient and the inhibitor or substrate concentration
(Cha, 1975). Both of these models can also produce inhibition of the
apparent initial velocity if the formation of the first EI complex
is rapid. Note that the noncompetitive model (Fig. 3B) predicts the
possibility of cooperativity for either inhibitor or substrate
in the steady state velocity (but not the initial) just as the
conformational form of the Ligand Induced Slow Transition model
(Fig. 1) does. The situation for threonine inhibition of homoserine
dehydrogenase (E. coli) that we described several years ago
(Bearer and Neet, 1978) is essentially this noncompetitive model,
except that it is complicated by a tetrameric structure and site-
site interactions of binding; the slow inhibition is on the order
of milliseconds and the kinetic contribution is in addition to
the equilibrium cooperativity.

Slow activation

     Slow activation is an analogous molecular process to that
described for inhibition, except that the modifier in Fig. 3B would
produce an activation, i.e. the primed cycle would be more active
than the unprimed (bottom) cycle. Laidler (Hijazi and Laidler, 1973)
has provided the steady state and pre-steady state equations for
at least one case of this mechanism. Numerous allosteric enzymes
have been described in which the activator produces its effects
slowly, e.g. the response of liver Acetyl CoA carboxylase to citrate
activation has been reported to require several minutes (Greenspan
and Lowenstein, 1968) and AMP (activator) affects the slow association and degree of activity of threonine dehydrase of E. coli
(Dunne and Wood, 1975).

     Hysteretic, allosteric activators are well known, but a
less well understood situation is one which involves hysteretic,
isosteric activators. The simplest (monomeric, one substrate enzyme) case of an essential, slow activator (Fig. 4C) could refer
either to an allosteric activator or to one required for the catalytic mechanism. Of particular interest are those essential activators that are cofactors of the reaction. The vitamin derived cofactor, thiamin pyrophosphate, TPP, appears to play this role in at
least four different enzymes. A slow transient lag occurs in the
activation by TPP of alpha-ketoglutarate dehydrogenase of cauli-
flower (Craig and Wedding, 1980), of pyruvate dehydrogenase of
E. coli (Graupe et al, 1982; Horn and Biswanger, 1983), of yeast
pyruvate decarboxylase (Hubner et al, 1978), and yeast transketolase
(Egan and Sable, 1981). The cofactor also promotes a dimerization
of the latter enzyme that does not appear to be at equilibrium
(Egan and Sable , 1981). In collaboration with Shreve and Sable

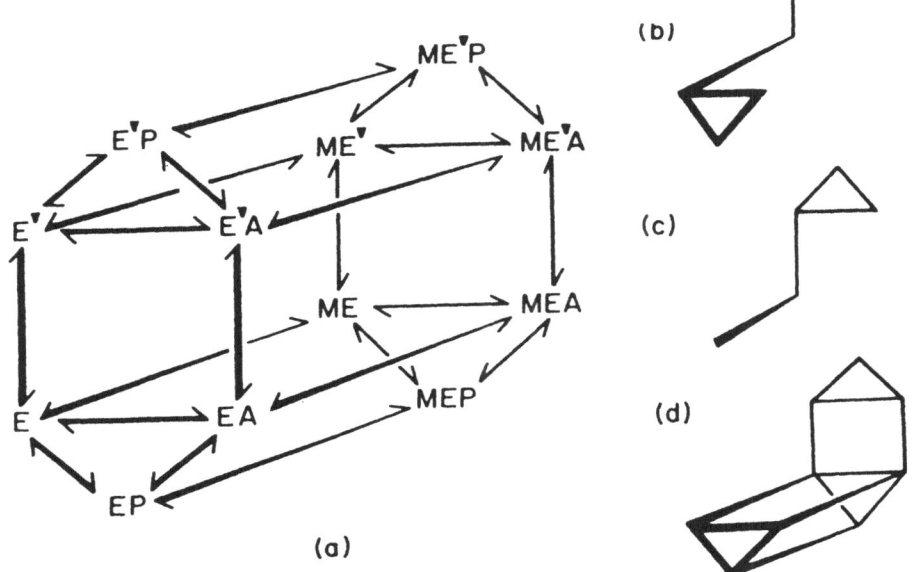

Fig. 4. (a) The general modifier effect on the Ligand Induced
Slow Transition (isomerization) Mechanism for a monomeric,
single substrate enzyme. The letters have the same meaning as Fig.1
with M representing an allosteric modifier. Enzyme forms with altered kinetic properties and altered distributions are demonstrated by the back plane (with M bound) of the mechanism.(b) A diagrammatic portion of the general modifier mechanism of (a) showing modifier binding only to free enzyme, E. In this case only competitive
inhibition can occur; compare Fig. 3a.(c) A diagrammatic portion
of the general modifier mechanism of (a) showing an essential
modifier binding to free enzyme, E, to induce the active conformation, E', through a slow isomerization.(d) A diagrammatic portion
of the general modifier mechanism of (a) showing modifier bound to
E, EA and EP but with the second, primed conformation only present
after induction by the modifier through a slow isomerization.

we have shown that the activation of transketolase is a complex
process that is not simply a slow binding of the ligand, as it
appears to be with α-ketoglutarate dehydrogenase (Craig and
Wedding, 1980), since the dependence of the reciprocal half-time
on TPP concentration is nonlinear for transketolase (Shreve et
al, 1984). Furthermore, the slow step can not simply be attributed

to the dimerization step itself, since this also gives the wrong
dependency of the rate on the TPP concentration. Whether this transition is related to the negative cooperativity observed in the
kinetic activation of the enzyme is as yet unclear. The fact that
four enzymes show slow activation by the TPP cofactor suggests
that this behavior must be important for its functioning or that
it is a necessary consequence of the reaction. It would be highly
unlikely that the slow process would be involved in a hysteretic
damping of the response to cellular concentrations of TPP,since
the latter are not known to change rapidly under different physiological conditions. A relationship to the cooperativity, as suggested
for two of the enzymes (Egan and Sable, 1981; Shreve et al, 1984;
Horn and Bisswanger, 1983), would seem to be the most likely possibility.

The general form of modifier effects on the slow transition
mechanism (Fig. 4A) is rather complex and, as usual, can account
for nearly any observation. The special case in Fig. 4B is identical
to the simple inhibition case (Fig. 3A) whereas the limiting scheme
in Fig. 4D is essentially the same as the inhibition of Fig. 3B. The
latter (Fig. 4D) can nicely accommodate slow effects seen in partial
competitive inhibition, noncompetitive inhibition, V or K type
activators, and effects of either type of modifier on apparent
initial velocities as well as steady state rates. The complete modifier (Fig. 4A), or the partial (Fig. 4D), slow transition mechanism (monomer) can give rise to terms with powers greater than two
in the rate equation for either substrate (A) or modifier (M) if the
binding of modifier itself (front to back steps) is not at rapid
 equilibrium. The full mechanism (Fig.4A) is necessary to explain
cases in which inhibition (or activation) does not cause changes
in cooperativity in the absence of modifier and hence represents
an entire new set of rate constants with similar relationships,
i.e. EA forms are not pulled. We have interpreted the partial
competitive inhibition of glucokinase by palmitoyl-CoA that does
not change the positive cooperativity with glucose in this fashion
(Tippett and Neet, 1982).

Slow membrane processes

We will now diverge for a moment from the main thrust of this
book, namely enzymes themselves, and briefly consider a related
topic, slow transition in membrane systems. In some cases these
may be enzymatic activities that are part of an integral membrane
protein, involved in transport or signalling. In other cases, the
slow processes may not have yet been associated with a classical
enzymatic activity but simply be manifested in binding equilibria.
Membranes may well be the archetypal systems for such slow responses since more restrictive forces on the diffusion, aggregation,
and/or conformational mobility may be operative (Ricard et al, 1984;
Hervagault et al, 1984).

Several notable examples have been reported in which hysteretic responses of membrane enzymes occur. The activation of rat liver plasma membrane adenylyl cyclase by GTP (or analogs) has been reported to be hysteretic, with half-times on the order of 2-10 minutes, and to be modulated by Mg^{2+} ion concentration (Rendell et al, 1975; Iyengar, 1981). Complex kinetic models, with a slow isomerization central to the argument, have been proposed to account for this behavior. Similarly, the vasopressin stimulated adenylyl cyclase of renal medulla has demonstrated both slow transients (Bockaert et al, 1973) and cooperative kinetics that are difficult to explain by classical equilibrium considerations. The Na^+, K^+ ATPase of renal plasma membrane has a slow process whose rate constant is a function of ATP and which has been interpreted as a slow isomerization between different kinetic forms (Cantley and Josephson, 1976). On the other hand, a transient in the kinetics of ion channel opening of the membranes or vesicles of excitable cells and neuromuscular junctions in response to acetylcholine has been reported by several laboratories (Changeux, 1981) and attributed to a mechanism involving a slow conformational change of the receptor. Whether these properties of membrane systems are integral to the physiological function (signalling or transport), contribute in some way to cooperative responses of the system, or merely reflect the sluggishness of highly viscous membrane systems remains to be seen in most instances.

SLOW INHIBITION OF YEAST HEXOKINASE BY METAL NUCLEOTIDES

The molecular basis for slow transitions in proteins. A case of 'induced misfit'

The analysis of the consequences of slow transitions have been reported in the whole family of hexokinases, including wheat germ hexokinase (Meunier et al, 1974; 1979) and glucokinase (Storer and Cornish-Bowden, 1977), representing a variety of kinds of transitions. The kinetics of the slow inhibition of the yeast hexokinase by a series of exchange inert, Me(III), or lanthanide metals have been extensively studied (Danenberg and Cleland, 1975; Viola et al, 1980; Peters and Neet, 1976; Neet et al, 1982). The conclusion from these studies is that the onset of full inhibition by complexes of ATP with such ions as Al(III), Cr(II), or Tb(III) only developes slowly and can be analyzed for the initial velocity, steady state velocity, and the half-time of the decay between them. Although the affinities are of the order of 1µM, these metal-nucleotides are not, in general, tight binding inhibitors since the enzyme concentration is well below this value. Our interest most recently has focussed on the structural changes underlying these kinetic transitions.

The crystallographic structure of hexokinase has shown (Bennett and Steitz, 1978; 1980; Steitz et al, 1981; Shoham and Steitz, 1982) in intricate beauty the bilobal nature of the enzyme and the closure of the cleft by certain sugars (glucose) but not others (N-substituted glucosamines, lyxose). The closure of the cleft is not the slow step or related to the kinetic transient, since fluorescent (Hogget and Kellett, 1976) and dye-linked (pH indicator) (Jentoft, Steuhr, and Neet, unpublished) temperature-jump,rapid kinetic studies of glucose binding have shown that the association of sugar with enzyme and the resultant conformational change is, not unexpectedly, complete within milliseconds. Thus, we have to look elsewhere for the structural change involved in the slow binding.

We (Ohning and Neet, 1983) have recently utilized the perturbation of fluorescence of bound TNS[*] to study the conformational states induced by a series of sugars, the energetics of metal-nucleotide binding, and the potential for slow transitions in various combinations of substrate-inhibitor pairs that go through the molecular transition associated with the slow inhibition. In addition, these changes may be putative analogs of the conformational changes during catalysis of a good substrate.

The inset to Fig. 5 shows the spectrum of yeast hexokinase in the presence and absence of TNS. A variety of catalytic and fluorescent studies have demonstrated that the TNS is not affecting the binding of substrates and vice versa . Furthermore, only one TNS molecule is bound per hexokinase monomer and it is not displaced by sugar and nucleotides under these experimental conditions so changes in fluorescence reflect environmental changes (Ohning and Neet, 1983). The spectra indicate that energy transfer between tryptophan and TNS occurs. Many of the experiments were also repeated using the intrinsic fluorescence of hexokinase and generally confirm the result with TNS.

The curve in Fig. 5 is a representative binding curve using the quenching of the TNS fluorescence caused by CrATP in the absence of sugars to determine the binding constant. The data of this type were utilized to determine the maximum change in fluorescence of TNS caused by sugar alone (Fig. 6, top) or by metal nucleotide in the presence or absence of sugar (not shown), and to determine the energetics of interaction determined from the binding constant (Fig. 6, bottom), presented as the change in free energy relative to that of the metal-nucleotide binding to hexokinase in the

----

[*] Abbreviations used are : TNS, 6-(p-toluidinyl)naphtalene-2-sulfonic acid; NGF, Nerve Growth Factor; PC12, pheochromocytoma 12 cell line; TPP, thiamine pyrophosphate.

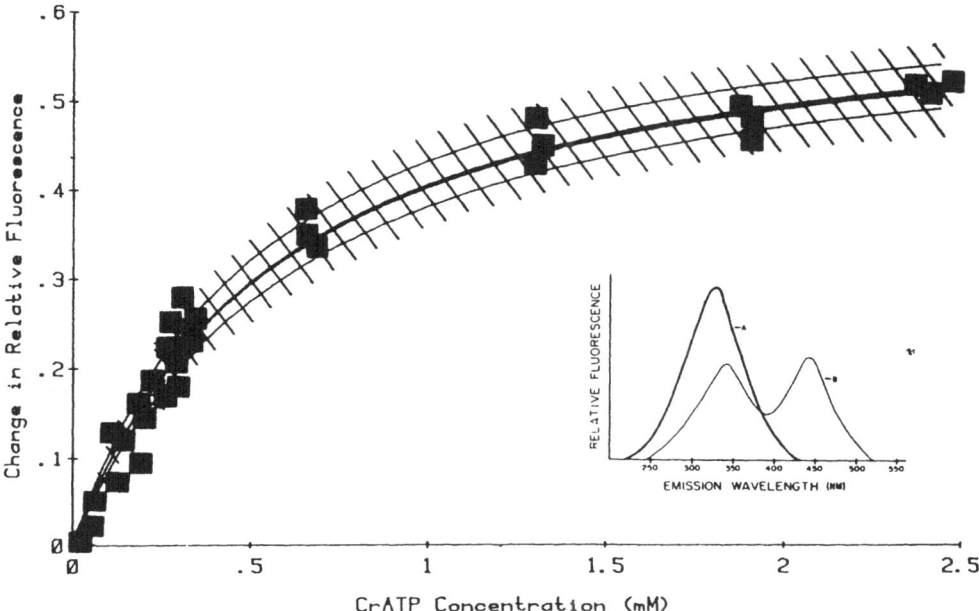

Fig. 5. Titration of yeast hexokinase with CrATP monitored by the quenching of fluorescence of the TNS probe by excitation at 295 nm and emission at 446 nm. The solid line is a non linear least square computer fit to the data with a value for the K_d of 0.58 mM. The light lines represent the standard error limits of the derived parameters and the crosshatched area is the 95% confidence limit. (inset) Emission spectrum (excitation at 295 nm) of hexokinase in 100 mM triethanolamine, pH 6.5 with (B) and without (A) 0.1mM TNS. (Reprinted by permission from Ohning and Neet , 1983).

absence of sugar ligand. Glucose and fructose gave the largest fluorescent enhancement (Fig. 6) , in agreement with their ability to close the cleft (Steitz et al, 1981) and act as good substrates (Viola et al, 1980). The other sugars could be distinguished on the basis of the degree of fluorescent change they allowed, which correlated well with their effectiveness as substrates or inhibitors.

The binding of nucleotides in the presence of sugars resulted in a quenching of fluorescence (relative to that in the presence of sugar alone) in all cases measured. Even for those sugars that did not cause a fluorescent change and either did (lyxose) or did not (Glc-NAc) promote subsequent metal nucleotide binding, the addition

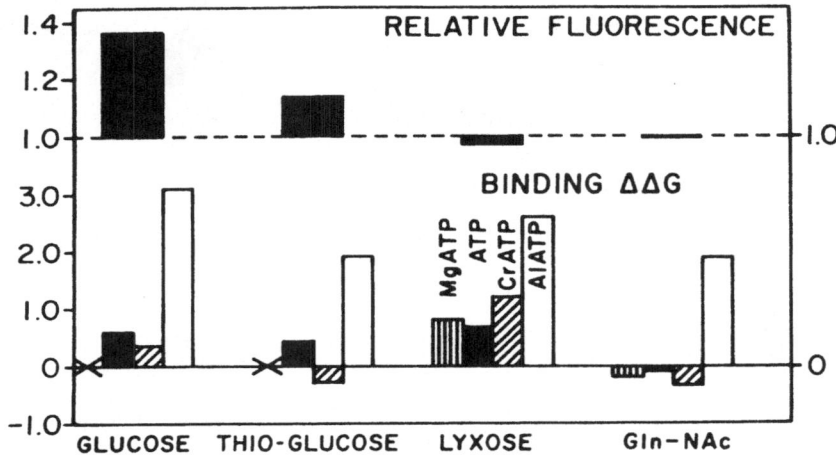

Fig. 6. Relative fluorescence (top) and change in the free energy change (bottom) of yeast hexokinase in the presence of four sugars and four metal-nucleotides. Relative fluorescence (top) is presented as the ratio of the fluorescence (arbitrary units) in the presence and absence of each sugar. The change in free energy (bottom), $\Delta\Delta G$(kcal/mol), is the difference between the ΔG for binding of the metal-nucleotide in the presence and absence of the sugar. For the nucleotides the symbols are : MgATP, vertical striped bar; ATP, solid bar; CrATP, slanted hatch-marked bar; and AlATP, open bar. The symbol (X) indicates that the measurements could not be made due to catalysis of an active complex (Data from Ohning and Neet, 1983).

of the nucleotides affected the fluorescence by quenching. Note, for example (Fig. 6), that lyxose , which does not enhance the fluorescence of bound TNS (top panel) nor close the cleft in crystals (Shoham and Steitz, 1982), still promotes the binding of all metal nucleotides tested (bottom panel), presumably through local alterations in the ATP-binding lobe. 5-Thio-glucose is a moderately weak substrate and has intermediate properties throughout.

The effect of each type of sugar on the subsequent binding of
metal-nucleotide (bottom panel of Fig.6) was specific for each metal
nucleotide but also reflected the sugar ligand used. As might be expected, little correlation appeared between the extent of fluorescent
quenching and the relative energy of binding. AlATP appears to have
large responses mainly because the binding of AlATP to free enzyme
essentially does not occur in the absence of sugars (the dissocation
constant is estimated here to be greater than 1mM). Thus, whether the
cleft remains open or not, nucleotide binding can occur with a perturbation of the environment of the TNS, probably due to a local conformational change. Neither sugars nor nucleotides perturbed the circular
dichroism spectrum.

We conclude from these results that each carbohydrate induces
a specific conformational change wich may also affect nucleotide
binding regardless of whether the cleft has closed. However, a
further conformational change, of a different qualitative nature,
must occur with metal nucleotide binding for catalysis to occur ...
a conclusion also reached by Steitz et al, 1981 (Bennett and
Steitz, 1980) from distances in the crystal between substrate binding
sites. We have not yet been able to draw a correlation between fluorescence quenching and the ability of the metal nucleotide to undergo
catalysis; this may be due to the fact that we have only studied
relatively good analogs of the true substrate that may induce the
active conformation but be inactive for other reasons, e.g. electronic interactions.

Once having ruled out the cleft closure and domain movement
with sugars as the slow transition, we were interested in determining if a slow fluorescence change could be observed in the presence of metal nucleotides and correlated with the slow inhibition
observed in assays. The only metal which gave observable slow
changes in TNS fluorescence was CrATP (Ohning, 1983). After an initial rapid quenching (36%)(used for the above binding calculations),
CrATP caused a further 16% quenching over the period of several
minutes. If glucose or lyxose was added at this point, an initally
rapid and then much slower enhancement of fluorescence was observed.
When the order of addition was reversed , CrATP added to a glucose-
hexokinase complex produced only a small quenching followed by a
small, slow enhancement of the TNS emission at 446 nm. Due to the
small changes in the presence of sugars, the only time courses
that were capable of being extensively analyzed were those, initially mentioned, of CrATP with hexokinase and no sugar.

The slow fluorescent effect of CrATP could only be observed
over a narrow range of concentrations and was interpreted as
reflecting the fluorescence of CrATP bound to both states of hexokinase that effectively saturate at different concentrations (Fig.
7). The crosshatched area of this graph indicates the extent and

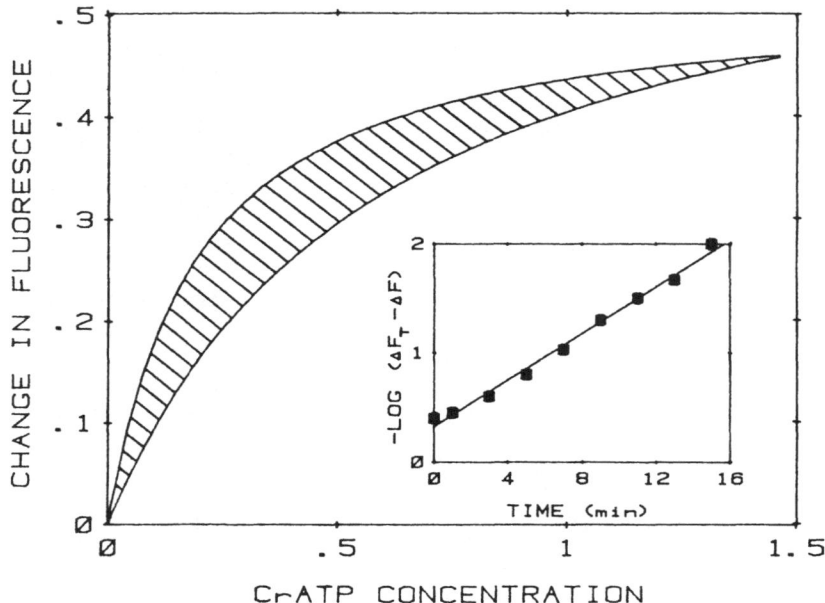

CrATP CONCENTRATION

Fig. 7. Binding of CrATP to yeast hexokinase monitored by quenching of TNS fluorescence. The curves are a diagrammatic representation of the K_d for the initial (1 min) and the final (20 min) binding of CrATP. The upper line is drawn for a K_d of 0.19 mM, a Δ fluorescence maximum of 0.52, and represents the binding parameters of CrATP after the transient. The lower line is drawn for a K_d of 0.58 mM, a Δ fluorescence maximum of 0.64, and represents the initial binding of CrATP. The cross-hatched area is the extent and accessible range of observation of the transient by this method. (Inset) First order plot of the quenching of the TNS fluorescence by 0.33 mM CrATP. Hexokinase concentration was 0.1 mg/ml in 100 mM triethanolamine, pH 6.5. $(\Delta FT - \Delta F)$ is the difference between the maximum fluorescence quenching (ΔFT) and the quenching (ΔF) at anytime. The slope of the line is 0.16 min^{-1}. Excitation was at 295 nm and emission at 446 nm.

the concentration region accessible for observation of the second, slow transition. The fluorescence change at the optimal difference of the curves (0.33 mM CrATP) is first order with time (Fig. 7, inset). The rate constant for this process was invariant at 0.15 ± 0.06 min^{-1} over the accessible concentration range (0.1 to 0.5 mM).

Table 1. Kinetic parameters of the hexokinase slow transition.

	Fluorescence[a]		Kinetics[b]	
K_1	580 μM			
$K_{12/34}$	190 μM			
K_{1G}	320 μM	K_C	315 μM	4 μM
k_3/k_4	2.05	k_7/k_8	14.7	0.44
k_3+k_4	0.15 min^{-1}	k_7+k_8	2.63 min^{-1}	2.3 min^{-1}
k_3	0.10 min^{-1}	k_7	2.5 min^{-1}	0.7 min^{-1}

Thus the observed slow fluorescence change has the correct properties to correspond to the slow inhibition (Table 1, left column). (a) K_1 and K_{1G} are the equilibrium constants for the initial change in TNS fluorescence in the absence and presence of glucose, respectively. $K_{12/34}$ is the overall equilibrium constant after the fluorescent transient. The ratio of k_3 to k_4 was obtained from these initial and final equilibrium constants without glucose. ($k_3 + k_4$) was the experimental value of $1/\tau$ of the fluorescence transient. k_3 was calculated. (b) $K_C = k_6/k_5$. k_5, k_6, k_7 and k_8 obtained from fit to the kinetic inhibition data with the value of K_C fixed at 315 or 4 uM.

The transient in the assay during the slow inhibition process has been reported (Danenberg and Cleland, 1975; Peters and Neet, 1976). We present here (Fig. 8) the dependence of the apparent rate constant (τ^{-1}) on CrATP concentration. The rate constants were obtained by a non-linear least square fit to a first order decay process of 50 to 100 points automatically acquired over 10 to 20 minutes (Ohning, 1983). The leveling off of the curve is consistent with an isomerization that becomes rate limiting at high inhibitor concentration. Under these experimental conditions the binding of CrATP occurs predominantly with the enzyme-glucose complex, in contrast to the analysis of the fluorescent perturbation in the absence of sugar. These transient kinetic data have been fitted to the appropriate equation derived from the slow binding and inhibition (Cha, 1975). The minimal pre-steady state rate equation has four terms : k_5 and k_6 , the rate constants for association and dissociation, respectively, for CrATP and the initial encounter complex; and k_7 and k_8, the forward and reverse rate constants for the slow isomerization step. K_C is the equilibrium dissociation constant or ratio of k_6 to k_5 (Fig. 9). These are not

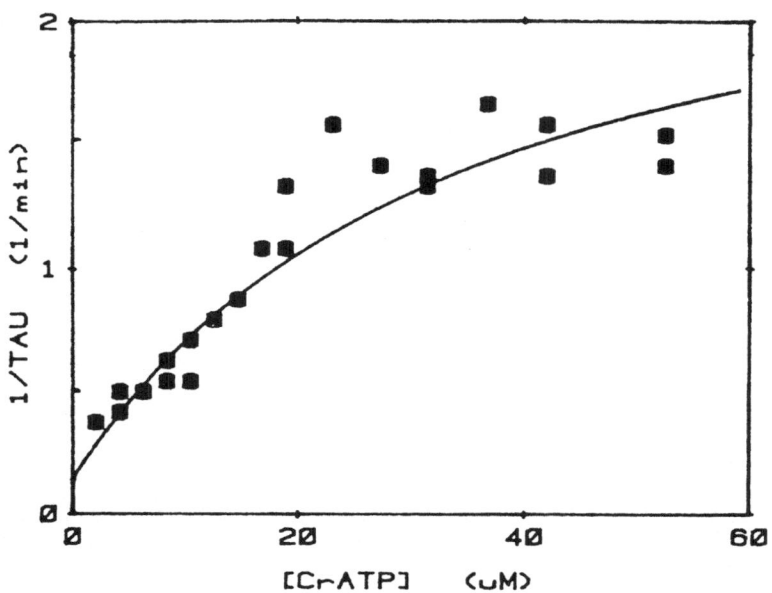

Fig. 8. The dependence of the reciprocal of the half-time $(1/\tau)$ for the slow inhibition of yeast hexokinase on the CrATP concentration. The values of $1/\tau$ were determined from a non-linear least square computer fit to progress curve with a first order decay (Neet and Ainslie, 1980). Assays were at pH 6.5 , 100 mM triethanolamine, with 2 mM ATP, 20 mM glucose and 10 mM MgCl$_2$.

independently variable constants in the equation so that one or more have to be fixed to fit the data to the remaining constants.

(1) $\tau^{-1} = \left[k_6 k_8 + k_5 [C] (k_7 + k_8)(1 + [M]/K_m)\right]/\left[k_7 + k_8 + k_5 [C](1 + [M]/K_m)\right]$

where : k_5, k_6, k_7, k_8 are the individual rate constants; $[C]$ and $[M]$ are the concentrations of CrATP and MgATP, respectively; K_m is the Michaelis constant for MgATP.

The line in Fig. 8 is drawn for a fixed value of K_c, 0.32 mM, in agreement with the K_1 determined by initial fluorescence titration of the hexokinase-glucose complex. The other fitted parameters are shown in Table 1 (first column under kinetics) and give a reasonable fit to the data. From this comparison to the rates deter-

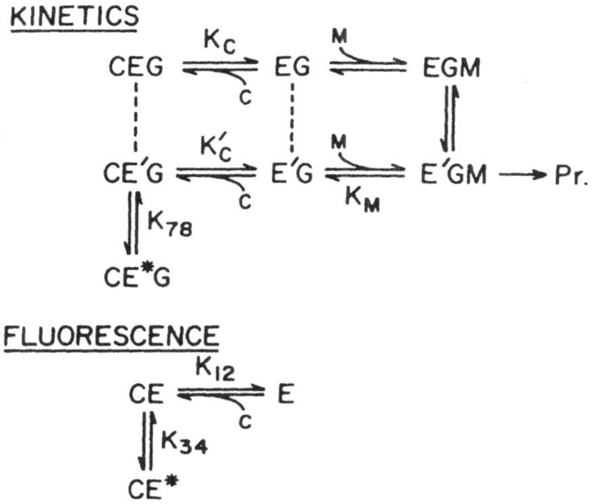

KINETICS

FLUORESCENCE

Fig.9. Model of the observed slow transitions in yeast hexokinase. The top line represents the transient observed in the slow inhibition of the enzymatic assay. The bottom line represents the transient observed by quenching of TNS fluorescence. The letters are used as follows : $\underline{E}$, $\underline{E}'$ and $\underline{E}^*$ - three conformations of hexokinase; $\underline{M}$ - MgATP; $\underline{C}$ - CrATP; $\underline{G}$ - glucose; $\underline{Pr.}$ product formed. K_{12}, K_C and $\overline{K}'_C$ are the dissociation constants for CrATP from $\underline{E}$, $\underline{EG}$ and $\underline{E'G}$, respectively. $K_{34}(= k_3/k_4)$ and $K_{78}(=k_7/k_8)$ are the equilibrium constants for the slow isomerization step of the fluorescence quenching and enzymatic assay, respectively. Vertical lines indicate interconversions between conformational forms which, in the kinetics, occur more readily for the active, ternary $\underline{EGM}$ complex.

mined by fluorescence (Table 1, first column), the rate of isomerization of the hexokinase-glucose-CrATP complex appears to be about 25 times faster than the rate of isomerization of the hexokinase-CrATP complex. Furthermore, the equilibrium isomerization constant appears to be about 7 fold larger in the presence of glucose. If this interpretation is correct, the closing of the cleft promotes the slow step both kinetically and thermodynamically.

Also shown in Table 1 is another adequate fit which was obtained by fixing the value of K_c (the dissociation constant of the initial CrATP complex) at the value of the inhibition constant (K_i) of the initial velocity (last column, Table 1). The rate constant of isomerization in this case is only 7 times larger than in the absence of glucose, but the equilibrium constant is actually smaller (5 fold). Our current interpretation to account for all these observations includes (Fig. 9) an initial interaction of CrATP with the hexokinase-glucose complex (320 μM), a conformationally different hexokinase which is induced by ATP, $\underline{E'G}$, and a kinetically significant binding (4 μM) of CrATP with this form of the enzyme. Note the similarity of this figure to that of the generalized inhibition of Fig. 3B or Fig. 4A. Either calculated constant appears to be a valid estimate and can give rise to the final, observed steady state inhibition constant, after 'tightening', of about 1 μM. The TNS fluorescence signal appears, indeed, to be measuring a physical parameter of the kinetically important slow transition of the slow inhibition.

Returning to the question of the molecular nature of the slow transition, we can say several things. Association-dissociation reactions are not involved since, under the experimental conditions used, the enzyme exists solely as a monomeric species (Furman and Neet, 1983). The domain movement with good sugars is rapid but occurs mainly at the 'hinge' region. The subsequent binding of CrATP also occurs without significant change in secondary structure, but with pronounced effects on the environment of the TNS reporter molecule. The magnitude of the structural change is uncertain; however, the magnitude of the slow fluorescence quenching by CrATP is about one-third that of the rapid quenching. The slow change can occur in the absence of sugar, although even more slowly, suggesting that it represents a local change in one of the lobes of hexokinase and not directly inter-domain interactions. However, both the rate and the equilibrium position are strongly influenced by sugar binding and domain closure.

The large protein structural changes previously suggested as possible slow transitions (Neet and Ainslie, 1980) do not appear to be as likely candidates now, and we may well have to look for much more subtle conformational changes with unusually high energy barriers. Yeast hexokinase remains an exciting and productive enzyme for obtaining details of the structural basis for the slow isomerizations based upon the crystallographic structure and a model for related kinases and other more distant proteins.

SLOW BINDING OF NERVE GROWTH FACTOR TO MEMBRANE RECEPTORS

The interaction of Nerve Growth Factor (NGF) with its membrane receptor demonstates some aspects of hysteresis that may provide important insights into the action of this (paracrine) hormone-like protein. NGF stimulates sensory and sympathetic neurons to produce neurites (axons), but in the process has pleiotropic actions on the responsive cells. The many responses are mediated by a plasma membrane receptor, but the nature of the second messenger(s), if any, is still under debate (Greene and Schooter, 1980; Vinores and Guroff, 1980). The aspect that will be focussed on here is the slow developement of the binding of radiolabeled NGF to the receptor on the surface of a responsive, model cell line, the pheochromo-cytoma (PC12) cells (Green and Tischler, 1976). The binding in these cells is characterized by two affinities of about 50 and 500 pM with 150,000 binding sites per cell (Woodruff, 1983). In many ways the structure of the hormone, the cellular binding, and the general intracellular responses to NGF are similar to the better known hormone, insulin, which acts on different tissues with different specific responses.

The dissociation of cellular bound NGF is characterized by two exponential decays, when displaced by either excess unlabelled NGF or by a second subunit of the oligomeric complex, corresponding to dissociation constants of 5×10^{-2} s^{-1} and 7×10^{-4} s^{-1}. The process that we are interested in here is demonstrated by the association of NGF with cells over about 30 min (Fig. 10). In high affinity, bimolecular systems such as this, it is important to determine that the 'slow' binding is not simply due to the low concentration used for experimental conditions. The raw binding data are shown on the left (Fig. 10), analyses for a simple bimolecular reaction on the upper right, and analysis for a reversible bimolecular reaction on the lower right. The lack of linearity indicates that the system can not be represented by a simple binding to a single type of receptor, but suggests the presence of either two receptors, a fast and a slow, or conversion (perhaps isomerization) between receptor forms. The faster rate constant estimated from these data is about 1×10^7 M^{-1}s^{-1} and the slower is at least 10 fold lower.

A different way of examining this process is by analysis for the slow and the fast receptors by their difference in ability to be dissociated at low temperatures (Landreth and Shooter, 1980; Schechter and Bothwell, 1981). Cells were incubated with ^{125}I-labelled NGF (525 pM) for 4 min, and one aliquot washed to remove unbound NGF. After various times of further incubation, the cells were examined for the slow receptors by displacing the low affinity receptors with an excess of unlabelled NGF at 4° (Fig. 11). The total binding remained constant, compared to the upper line for unwashed cells. The slow (or high affinity) receptors however, tend to continue to increase. Experiments analogous to this have been interpreted by Landreth and Shooter (1980) to indicate a conversion of fast (low affinity) receptors to slow (high affinity) receptors.

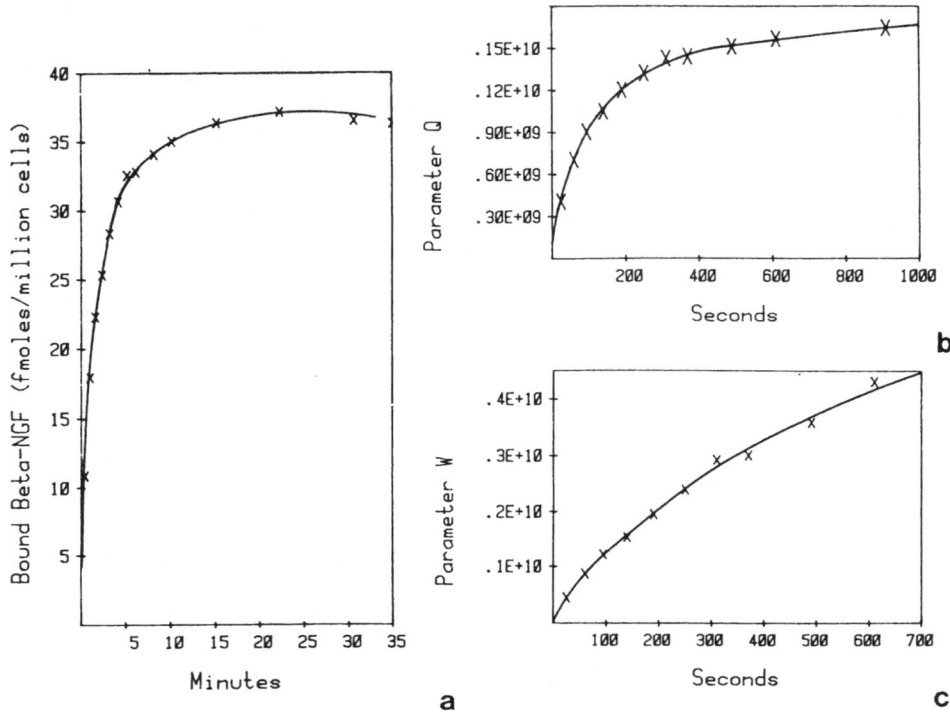

Fig. 10. Rate of association of the binding of β NGF to PC12 cell
line. 112 pM ^{125}I-β NGF was incubated with 1.6 x 10^6 cells per ml
for the indicated times at 37° and binding measured by sedimenta-
tion through a one step sucrose gradient (Landreth and Shooter,
1980). Specific binding was determined as the difference between
total binding and non-specific binding measured in a parallel
incubation with 1000 fold excess of unlabeled NGF. (a, left)Speci-
fically bound NGF vs time of incubation. (b, upper right) Plot for
an irreversible second binding process.
$$Q = k_1 t = (a-b)^{-1} \ln \left[(b(a-x))/((a)(b-x)) \right]$$
(c, lower right) Plot for a reversible second order binding process.
$$W = k_1 t = Z^{-1} \ln \left[((a+b+K_{eq}-Z)(2x-a-b-K_{eq}-Z))/((a+b+K_{eq}+Z) \; X \right.$$
$$\left. (2x-a-b-K_{eq}+Z)) \right]$$
where k_1 = second order rate constant; t = time; a = initial NGF
concentration; b = initial receptor concentration; x=bound NGF
concentration at any time; K_{eq} = apparent equilibrium dissociation constant; $Z=\left[(-a-b-K_{eq})^2 - 4ab \right]^{1/2}$.

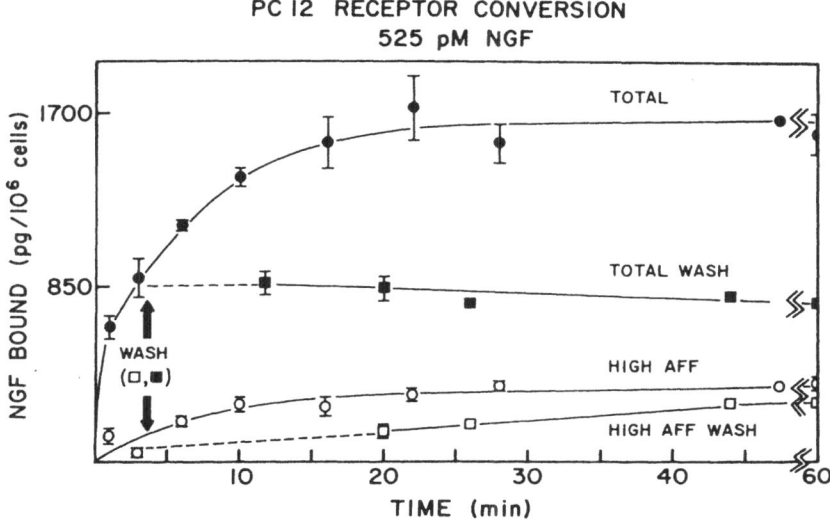

Fig. 11. Conversion of NGF receptors. 525 pM ^{125}I-NGF was incubated
with 1 x 10^6 PC12 cells per ml at 37°. Total specific binding was
measured at the indicated times by the one step sucrose sedimenta-
tion method (solid circles) and tight binding was measured by a
cold chase with 1000 fold excess of unlabeled NGF at 0° for 30 min
(open cicles). At 4 min (arrow) an aliquot of cells was washed
with buffer and resuspended in the absence of NGF; total specific
(solid squares) and tight (open squares) binding was measured
on these cells, as described above, at indicated times.

Other interpretations are possible, including diffusional barriers
and subsequent rebinding of dissociated NGF (Schechter and Bothwell,
1981). However, we have been able to reproduce much of the slow
and fast receptor kinetics utilizing the α subunit of NGF to bind
the β NGF and promote dissociation, suggesting that the kinetics
truly represents differently responding receptor populations. Conversion of low affinity to high affinity binding can also occur by
other means (Block and Bothwell, 1983), including incubation of
the cells with a lectin, wheat germ agglutinin, which causes a low
to high affinity receptor conversion over about 10 minutes (Buxser
et al, 1983).

Two models incorporating these slow binding properties have

been postulated and are special cases of a general model
(Fig. 12). The sequential conversion model (Landreth and Schooter,
1980) would involve binding (top) followed by slow isomerization
(right) to βR'. The parallel receptor model would involve two
classes of pre-existing receptors, a fast (R) or slow (R') binding.
These receptors might be independent or undergo interconversion
(βR to βR') in the complete model. In comparison, several other
hormone binding systems have been hypothesized to follow a similar
cyclic mechanism such as that of Fig. 12 . For the NGF receptor
we can not yet define the molecular basis of the several steps,
but, regardless of the detailed pathway, the implications of the
slow steps in the process raise some interesting questions of
function.

Three general points may be made concerning the possibility
of the contribution of slow transitions to membrane systems as
exemplified by the NGF system. (a) Slow (hysteretic) properties
may be quite common in membranes where important events include two-
dimensional diffusion (patching, capping, microcluster formation),
diffusion through a liquid crystalline array of phospholipid chains,
transmembrane generation of signals, transport of small molecules,
rotational diffusion of integral membrane proteins, and flip-flop
of phospholipids. (b) The multitude of pleiotropic responses to NGF
(or other hormones such as insulin) are not adequately explained
by a single second messenger, but might be more easily interpreted
if the multiple states of the receptor complex, i.e. βR and βR',
contribute different signals to the intracellular milieu. For NGF,
the ultimate responses of neurons (that occur at different times)
include amino acid and ion transport into the cell, specific enzyme induction, stimulation of protein synthesis, protein phosphorylation, cell surface morphology changes, adhesiveness changes,
microtubule reorganization, and neurite extension (Greene and
Shooter, 1980; Vinores and Guroff, 1980). (c) Finally, the temporal
course of the hysteretic mechanism (Fig. 12) for NGF may be important. When the conversion occurs, then the βR (upper right) might
cause timed signals to initiate certain responses, whereas the
βR' (lower right) which develops more slowly, may produce other,
later signals. In this context the experimentally observed time
course over 20-30 minutes to attain steady state binding must be
contrasted with the 3-4 hours that NGF must remain in contact with
cells for commitment to stimulation and with the 1 to 6 days
necessary for full neurite response (depending upon conditions).
Certainly, other reactions (e.g. biosynthesis of intermediates)
are occurring within the cell that account for the longest time
lags. However, hysteretic responses in membrane signalling systems
may provide an additional way for a single protein-protein interaction to help coordinate the highly complex interactions of
multiple signals in order to integrate, temporally, the multiple
responses within a cell. The correct overall biological response
is then produced at the appropriate time.

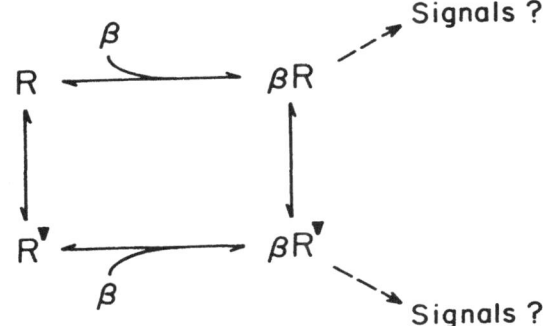

Fig. 12. Hysteretic model for the binding and response of cellular receptors with βNGF. β represents the NGF; R and R' two conformations of the plasma membrane receptor on PC12 or neuronal cells. Verticle lines represent slow interconversion in the presence or, potentially, in the absence of NGF. 'Signals' represents the variety of internal events in response to NGF stimulation (Greene and Shooter, 1980).

CONCLUSION AND OVERVIEW

Finally , a brief discussion of the state of hysteretic enzymes reported in the literature is appropriate. Over 75 reports of various types of hysteretic behavior have been demonstrated in the past 15 years. Comprehensive lists have been recently published (Neet and Ainslie,1980; Kurganov, 1982). A classification scheme has been suggested (Neet and Ainslie, 1980) based upon the particular contribution of the hysteresis to the enzyme's function. Although it is not always easy to assign this raison d'être , some generalizations can be made. The majority (about 60%) of enzymes studied have been suggested to belong to either Class II, slow damping response to physiological changes in ligand concentration, or class IV, contributing to the cooperativity of an oligomeric enzyme. Distinction between these two classes is often difficult without extensive work, both in vitro and in vivo . Some enzymes appear to have transients for trivial reasons (Class I) and a few appear to cause cooperativity in a monomeric enzyme (Class III). Future work in this area will necessarily require much attention to refining and defining these roles of hysteresis in enzyme function and will require detailed study of many interesting mechanisms.

REFERENCES

Ainslie, G.R., Shill, J.P. and Neet, K.E. (1972) J. Biol. Chem.247, 7088-7096.
Bearer, C.F. and Neet, K.E. (1978) Biochemistry 17, 3512-3516; 3517-3522.
Bennett, W,S. and Steitz, T.A.(1980) J. Mol. Biol. 140, 183-209 ; 211-230.
Block, T. and Bothwell, M.A. (1983) J. Neurochem. 40, 1654-1663.
Bockaert, J., Roy, C. , Rajerison, R. and Jard, S. (1973) J. Biol. Chem. 248,5922-5931.
Buxser, S.E., Kelleher, D.J., Watson, L., Puma, P. and Johnson, G.L. (1983) J. Biol. Chem. 258 , 3741-3749.
Cantley, L.J.Jr. and Josephson, L. (1976) Biochemistry 15, 5280-5286.
Changeux, J-P. (1981) The Harvey Lectures 75, 85-254.
Cha, S. (1975) Biochem. Pharm. 24, 2177-2185, 25, 1561.
Craig, D.W. and Wedding, R.T. (1980) J. Biol. Chem. 255, 5769-5775.
Danenberg, D.D. and Cleland, W.W. (1975) Biochemistry 14, 28-39.
Dunne, C.P. and Wood, W.A. (1975) Curr. Topics Cell. Reg. 9, 65-101.
Egan, R. and Sable, H.Z. (1981) J. Biol. Chem. 256, 4877-4883.
Frieden, C. (1970) J. Biol. Chem. 245 , 5788-5799.
Frieden, C. (1979) Ann. Rev. Biochem. 48, 471-489.
Furman, T.C. and Neet, K.E. (1983) J. Biol. Chem. 258, 4930-4936.
Graupe, K., Abusaud, M., Karfunkel, H. and Bisswanger, H. (1982) Biochemistry 21, 1386-1394,
Greene, L.A. and Shooter, E.M. (1980) Ann. Rev. Neurosci. 3 , 353-402.

Greene, L.A. and Tischler, A.T. (1976) Proc.Natl. Acad. Sci (USA)
73 , 2424-2428.
Greenspan, M.D. and Lowenstein, J.M. (1968) J. Biol. Chem. 243 ,
6273-6278.
Gregoriou, M., Trayer, I. and Cornish-Bowden, A. (1981) Biochemistry,
20, 499-506.
Hervagault, J.F., Breton, J., Kernevez, J.P. , Rajani, J. and
Thomas, D. (1984) , This Volume.
Hess, G.P., Cash, D.J. and Aoshima, H. (1979) Nature 282,329-331.
Hijazi, N.H. and Laidler, K.J. (1973) Can. J. Biochem. 51 , 806-814.
Hoggett, J.G. and Kellett, G.L. (1976) Eur. J. Biochem. 68 , 347-
353.
Horn, F. and Bisswanger, H. (1983) J. Biol. Chem. 258, 6912-6919.
Hubner, G., Weidhase, R. and Schellenberger, A. (1978) Eur. J.
Biochem. 92 , 175-181.
Iyengar, R. (1981) J. Biol. Chem. 256, 11042-11050.
Jarabak, R. and Westley, J. (1974) Biochemistry 13 , 3233-3236 ; 3237-
3239.
Keleti, T. (1968) Acta Biochim. Biophys. Acad. Sci. Hung. 3 , 247-
258.
Klinov, S.V. and Kurganov, B.I. (1982) J. Theor. Biol. 98,73-90.
Kurganov, B.I. (1977) J. Theor. Biol. 68 , 521-543.
Kurganov, B.I. (1982)'Allosteric Enzymes', John Wiley and Sons, New
York.
Landreth, G.E. and Schooter, E.M. (1980) Proc. Natl. Acad. Sci.(USA)
4751-4755.
Meunier, J.C., Buc, J. and Ricard, J. (1974) Eur. J. Biochem. 49,
209-223.
Meunier, J.C., Buc, J. and Ricard, J. (1979) Eur. J. Biochem. 97,
573-583.
Morrison, J.F. (1982) Trends Biochem. Sci. 7, 102-105.
Neet, K.E. (1979) Bull. Biol. Med. 4 , 101-138.
Neet, K.E. and Ainslie, G.R. (1980) Meth. Enzymol. 64, 192-226.
Neet, K.E., Furman, T.C. and Hueston, W.J. (1982) Arch . Biochem.
Biophys. 213 , 14-25.
Neubig, R.R. and Cohen, J.B. (1980) Biochemistry 19, 2770-2779.
Ohning, G.V. (1983) PhD Dissertation , Case Western Reserve
University.
Ohning, G.V. and Neet, K.E. (1983) Biochemistry 22 , 2986-2995.
Peters, B.A. and Neet, K.E. (1976) J. Biol. Chem. 251 , 7521-7525.
Rabin, B.R. (1967) Biochem. J. 102 , 22c.
Rendell, M. , Salomon, Y., Lin, M.C. , Rodbell, M. and Berman, M.
(1975) J. Biol. Chem. 250 , 4253-4260.
Ricard, J., Meunier, J.C. and Buc, J. (1974) Eur. J. Biochem. 49,
195-208.
Ricard, J., Noat, G. and Crasnier, M. (1984) This volume.
Schechter, A.L. and Bothwell, M.A. (1981) Cell 24 , 867-874.
Shoham, M. and Steitz, T.A. (1980) J. Mol. Biol. 140 , 1-14.

Shoham, M. and Steitz, T.A. (1982) Biochim. Biophys. Acta 705 , 380-384.

Shreve, D., Hagerty, J., Neet, K.E. and Sable, H.Z. (1984) in preparation.

Steitz, T.A., Shoham, M. and Bennett, W.S. (1981) Phil. Trans. R. Soc. Lond. 293B , 43-52.

Storer, A.C. and Cornish-Bowden, A. (1977) Biochem. J. 165, 61-69.

Tippett, P.S. and Neet, K.E. (1982) J. Biol. Chem. 257 , 12846-12852.

Vale, R.D. and Shooter, E.M. (1982) J. Cell. Biol. 94 , 710-717.

Vinores, S. and Guroff, G. (1980) Ann. Rev. Biophys. Bioeng. 9 , 223-257.

Viola, R.E., Morrison, J.F. and Cleland, W.W. (1980) Biochemistry 19 , 3131-3137.

Whitehead, E. (1970) Progress. Biophys. 21 , 321-397.

Whitehead, E. (1976) Biochem. J. 159 , 449-456.

Williams, J.W. and Morrison, J.F. (1979) Meth. Enzymol. 63 , 437-467.

Witzel, H. (1967) Hoppe Seyler's Z. Physiol. Chem. 348, 1249-1250.

Yankner, B.A. and Shooter, E.M. (1982) Ann. Rev. Biochem. 51, 845-868.

Woodruff, N.R. (1983) PhD Dissertation, Case Western Reserve University.

RAT-LIVER GLUCOKINASE AS A MNEMONICAL ENZYME

Athel Cornish-Bowden, Mary Gregoriou[1] and
Denise Pollard-Knight[2]

Department of Biochemistry, University of Birmingham
P.O. Box 363, Birmingham B15 2TT, England
[1] Present address : Laboratory of Molecular Biophysics
Department of Zoology, University of Oxford, South
Park Roads, Oxford OX1 3PS, England
[2] Present address : Department of Biochemistry, University
of California, Berkeley, California, 94720, USA

INTRODUCTION

Four different isoenzymes catalyzing the phosphorylation of glucose by MgATP have been described in the rat, and a similar iso-enzyme pattern is found in other mammals (Ureta, 1982). Although these are all isoenzymes of hexokinase (ATP : D-hexose 6-phospho-transferase, EC 2.7.1.1), the liver isoenzyme, hexokinase type IV or type D, which is generally known as glucokinase, shows striking differences in its kinetic behavior from the other isoenzymes. These differences relate to the fact that glucose phosphorylation in liver is primarily a homeostatic mechanism to maintain a constant blood-glucose concentration, whereas elsewhere it provides for the needs of tissues for metabolic energy. In particular, glucokinase displays positive co-operativity with respect to glucose, and is insensitive to inhibition by glucose 6-phosphate (Niemeyer et al., 1975 ; Storer and Cornish-Bowden, 1976).

The co-operativity of glucokinase is of particular interest, because the enzyme is a 48 kDa monomer with only one active site (Holroyde et al., 1976 ; Cárdenas et al., 1978, 1979 ; Connolly and Trayer, 1979). Thus the co-operativity must be purely kinetic in origin and cannot arise from subunit interactions, as in the classical models of co-operativity (Monod et al., 1965 ; Koshland et al., 1966).

Although kinetic co-operativity in monomeric enzymes has

29

been known as a theoretical possibility for a long time (Frieden, 1964 ; Ferdinand, 1966), few experimental examples have been recognized until recently. The model of Frieden (1964) proposes a second substrate binding site on the molecule, and can thus be excluded from consideration in the case of glucokinase, though it may apply to some other enzymes, such as ribonucleoside triphosphate reductase from <u>Lactobacillus leichmannii</u> (Panagou <u>et al.</u>, 1972) and pancreatic ribonuclease (Walker <u>et al.</u>, 1975, 1976). The model of Ferdinand (1966) assumes that deviations from Michaelis-Menten behavior arise from variations in the order of substrate binding as the substrate concentrations change. This model has not been demonstrated to apply to any enzyme, though it may well make a contribution to the co-operativity of some, including glucokinase.

More recently, ideas outlined by Rabin (1968) have been extended into the "slow-transition model" of Ainslie <u>et al.</u> (1972) and the "mnemonical model" of Ricard <u>et al.</u> (1974), both of which depend on the idea that the enzyme may exist in two or more states that interconvert slowly enough for them to be far from equilibrium during the catalytic reaction. They differ from other models of co-operativity in being strictly single-site models that could in principle be applied to single-substrate enzymes. They were originally applied to yeast hexokinase (Shill and Neet, 1975) and to wheat-germ hexokinase (Meunier <u>et al.</u>, 1974). More recently the mnemonical model has been proposed as an explanation of the kinetics of rat-liver glucokinase (Storer and Cornish-Bowden, 1978) and of molluscan octopine dehydrogenase (Monneuse-Doublet <u>et al.</u>, 1978).

Cárdenas <u>et al.</u> (1978) initially interpreted glucokinase co-operativity in terms of a model similar to that of Ferdinand (1966), but they subsequently found that this model was inadequate to explain all of the data and expanded it to include the possibility of multiple forms of free enzyme (Cárdenas <u>et al.</u>, 1979). Simulation of this extended model on an analog computer (Olaverría <u>et al.</u>, 1983), using a stochastic method developed by Olaverría (1973, 1980) showed that it was capable of accounting for glucokinase co-operativity as well as its kinetics with alternative substrates such as mannose and 2-deoxyglucose. Similarly, Gregoriou <u>et al.</u> (1981) found that the mnemonical model with a compulsory order of substrate binding was unable to account for the kinetics of isotope exchange at equilibrium, and extended that model to allow MgATP to bind before glucose in some circumstances. Thus although there are some remaining differences in emphasis there is now little disagreement about the essential features of the glucokinase mechanism.

In this chapter we discuss how two different uses of isotopes in enzyme kinetics have shed light on the glucokinase mechanism.

A more general review, including discussion of the stereochemistry
of the reaction, may be found elsewhere (Pollard-Knight and Cornish-
Bowden, 1982). Elsewhere in this volume Neet et al. (1984) discuss
other enzymes in which slow steps are implicated in the mechanism.

ISOTOPE EXCHANGE AND ISOTOPE EFFECTS

Isotope exchange and isotope effects are two entirely different ways of studying a mechanism, and as both are discussed in this
chapter in relation to the glucokinase mechanism it is important to
distinguish between them and to recognize that they have almost nothing in common beyond the use of isotopes.

In an isotope-exchange experiment the isotope is virtually
always radioactive and is used solely as a trace label. In order
that the isotopic label shall have no effect on the kinetics of
the reaction it is placed remote from the site of reaction, separated from any reacting atom by at least two bonds. The use of a radioactive label allows one to trace the fate of individual reactive
molecules even if the bulk chemical reaction is at equilibrium or is
proceeding in the opposite direction from the isotope exchange.
This type of study can provide much less ambiguous information about
the order of substrate binding and product release than is available
from other techniques. For the glukinase reaction, labeling with
^{14}C allows measurement of the flux between glucose and glucose 6-
phosphate or between ATP and ADP, whereas labeling with ^{32}P allows
measurement of the flux between ATP and glucose 6-phosphate. It
does not appear possible to measure the fourth flux, between glucose and ADP, but fortunately one can obtain as much information as
needed from measurements of the other three.

By contrast, an isotope effect is an alteration in kinetics
or equilibrium constant brought about by isotopic substitution.
For the effect to be measurable the substitution must result in
an appreciable change in mass of an atom at the site of reaction.
Thus in practice the substitution is nearly always of ^1H by ^2H or
^3H : substitution of an atom directly involved in the reaction
produces a primary isotope effect, whereas substitution very close
to the site of reaction produces a (much smaller) secondary isotope effect. The simplest type of isotope effect to measure (though
not necessarily to interpret) is a solvent isotope effect brought
about by carrying out the reaction in ^2H_2O or in mixtures of ^1H_2O
and ^2H_2O. In this chapter we shall only discuss solvent isotope
effects in pure ^2H_2O, though we have also examined isotope effects
on glucokinase in mixed solvents (Pollard-Knight and Cornish-Bowden, 1984).

ORDER OF SUBSTRATE BINDING

An important aspect of the mnemonical model as an explanation

of glucokinase co-operativity is that glucose must be capable of
binding to the enzyme before MgATP. This is because the ability
of MgATP to induce glucose co-operativity is assumed to derive
from a very rapid reaction between MgATP and the enzyme-glucose
complex that prevents equilibration of glucose binding. It was thus
important to obtain independent information about the order of substrate binding and product release.

Although product inhibition experiments are commonly used for
obtaining information of this kind, they have serious drawbacks in
any but the simplest mechanisms, because their interpretation is
complicated by any unusual mechanistic features, such as isomerization of the free enzyme or non-productive side reactions. Moreover, in a product-inhibition study one cannot examine the order of
substrate binding separately from the order of product release. Various kinds of isotope-exchange experiment avoid these difficulties:
measurements at chemical equilibrium with all four substrate concentrations varied in constant ratio (Wedler and Boyer, 1972) provide a very sensitive way of detecting the existence of minor pathways of reaction, though they do not show whether such pathways make a significant contribution to the flux under ordinary steady-
state conditions : the powerful though less well known flux-ratio
method (Britton, 1966) can, however, be used under any conditions
and does allow examination of the order of binding in the steady
state. It requires the simultaneous measurement of transfer of label
from a doubly labeled product to two substrates. Although the expressions for the rates of transfer are extremely complicated the ratio
between them is normally rather simple : if the two substrates can
bind to the enzyme in either order the flux ratio shows a hyperbolic dependence on each substrate concentration ; but if they bind
in a compulsory order the flux ratio shows a linear dependence on
the concentration of the substrate that binds second and is independent of the concentration of the substrate that binds first.
This last characteristic is a special case of a general feature of
the method, which is that the flux ratio is unaffected by any steps
that occur in the mechanism outside the part of it that is involved
in the transfer of label. This is a major advantage of the flux-
ratio approach, because it means that it is unaffected by such com-
plications as isomerization of the free enzyme and any non-productive steps in the mechanism.

For the glucokinase reaction, measurement of isotope exchange
at chemical equilibrium according to Wedler and Boyer (1972) showed
that over a 30-fold range of concentrations the rates of exchange
between ATP and ADP and between glucose and glucose 6-phosphate
increased to saturation, with the former exchange about 20 times
faster than the latter (Gregoriou et al., 1981). The difference in
rates suggests that MgATP and MgADP are the "inner" part of substrates, but the failure to detect any inhibition of the exchange

between glucose and glucose 6-phosphate at high concentrations suggests that an alternative order of reaction, with MgATP binding before glucose, can occur as a minor pathway. Unfortunately the unfavorable equilibrium constant for the reaction makes it impracticable to extend measurements to higher concentrations, as this would mean concentrations of glucose 6-phosphate greater than 0.12 M.

Glucokinase does not catalyze exchange between glucose and glucose 6-phosphate in the absence of ATP and ADP, or between ATP and ADP in the absence of glucose and glucose 6-phosphate. However, if labeled ADP is added to a reacting mixture of ATP and glucose in the presence of glucokinase, transfer of label from ADP to ATP can be detected, suggesting that the catalytic reaction proceeds through a complex between enzyme and glucose 6-phosphate that can be "trapped" by labeled ADP. Similar experiments with labeled glucose 6-phosphate failed to detect a corresponding complex between enzyme and ADP, and experiments in the reverse direction detected an enzyme-glucose complex but no enzyme-ATP complex.

These trapping experiments suggested that the glucokinase reaction proceeds according to a compulsory-order mechanism with glucose binding first and glucose 6-phosphate released last. However, they were essentially qualitative in nature and needed to be supplemented by more precise information. This was provided by the flux-ratio measurements in Fig. 1. Doubly labeled glucose 6-phosphate was added to reaction mixtures containing glucose and ATP and the rates of transfer of ^{14}C to glucose and ^{32}P to ATP were measured in the same reaction mixtures. The ratio of the two fluxes was then examined as a function of the two substrate concentrations. No dependence on the concentration of glucose can be detected, but there is a linear dependence on the concentration of MgATP, exactly as predicted by Britton (1966) for a compulsory-order mechanism in which glucose binds before MgATP.

A flux-ratio experiment under the conditions of Fig. 1 is unaffected by complexities in the order of product release. Although this is advantageous in the sense that it allows the order of substrate binding to be studied in isolation, it does of course mean that the experiment provides no information about the order of product release. To obtain this, similar experiments were carried out with doubly labeled ATP. These gave corresponding results (see Fig. 5 of Gregoriou et al., 1981) indicating a compulsory-order mechanism for the reverse reaction, with glucose 6-phosphate binding before MgADP.

In summary, the isotope-exchange experiments indicate that the glucokinase reaction proceeds predominantly by a mechanism in which glucose binds before MgATP and MgADP is released before glucose 6-phosphate. Although the experiments at chemical equilibrium

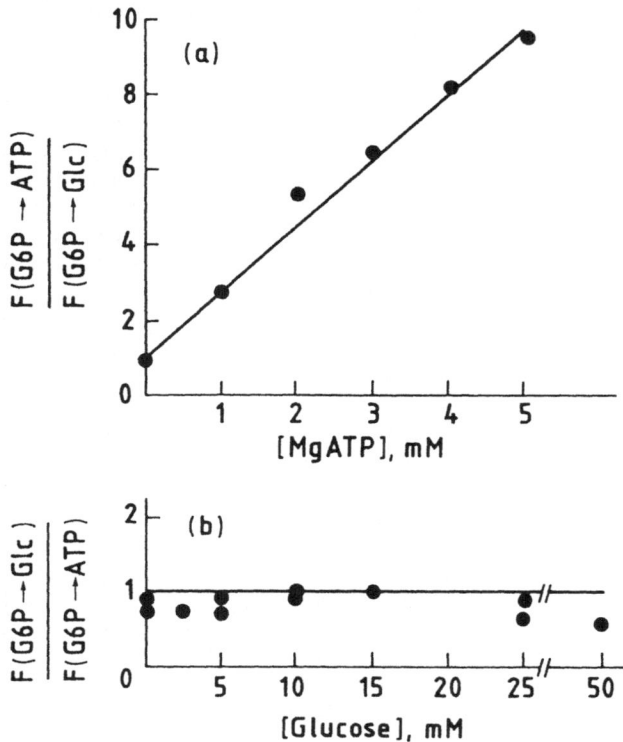

Figure 1. Flux-ratio measurements showing that glucose binds before ATP in the forward direction. The symbol $F(G6P \longrightarrow ATP)$ represents the flux from glucose 6-phosphate, determined by measurements of exchange of ^{32}P and $F(G6P \longrightarrow Glc)$ represents the corresponding flux from glucose 6-phosphate to glucose, measured with ^{14}C. (a) The flux ratio $F(G6P \longrightarrow ATP) / F(G6P \longrightarrow Glc)$ shows a linear dependence on the concentration of MgATP, but (b) there is no dependence on the concentration of glucose. According to the theory developed by Britton (1966), these results indicate a compulsory-order of binding of substrates with glucose binding before MgATP. The figure is reprinted with permission from Gregoriou et al. (1981) ; copyright 1981 American Chemical Society. Full experimental details may be found in the original.

indicated the possibility of an alternative pathway, this pathway appears to make no significant contribution to the flux under ordinary steady-state conditions. This therefore confirms one aspect of the mnemonical model as a mechanism for glucokinase co-operativity.

COMMENTS ON THE FLUX-RATIO METHOD

It is perhaps worth noting that the flux-ratio method is by no means as complicated, either theoretically or experimentally, as it may appear. Although a full derivation of the individual flux expressions leads to some complicated algebra, the underlying theory is fairly simple and can be understood without recourse to algebra (Cornish-Bowden and Gregoriou, 1981). The need for doubly labeled reactants also presents little problem in practice : there is no need for individual molecules to be doubly labeled (even if this were possible) ; it is quite sufficient to prepare singly labeled compounds and to mix them together for study.

SOLVENT ISOTOPE EFFECTS ON GLUCOKINASE

Various enzymes, such as pyruvate carboxylase (Irias et al., 1969), glutamate dehydrogenase (Henderson and Henderson, 1969), lactate dehydrogenase (Henderson et al., 1970) and fatty acid synthase (Smith, 1971), are more stable in ^2H_2O than in ^1H_2O. Although the mechanism of stabilization is not established, these and other observations suggest that ^2H_2O may stabilize hydrogen bonds, thus protecting enzymes against dissociation or denaturation. It would thus be expected to affect the equilibria between different conformational states of an enzyme, as major changes in conformation would be expected to be accompanied by redistribution of hydrogen bonds. We therefore studied the kinetics of glucokinase in ^2H_2O with a view to obtaining more information about the two postulated states of free enzyme in the mnemonical model.

Before trying to interpret changes of kinetics in ^2H_2O one should insure that there are no qualitative changes in pL dependence, where pL is a generalized pH that applies to ^2H_2O and to mixtures of ^1H_2O and ^2H_2O. In the case of glucokinase, and in contrast to results recently reported by Taylor et al. (1983) for yeast hexokinase, the shape of the pL profile is virtually unchanged in ^2H_2O, though the optimum is shifted upwards by about 0.5 pL unit : this shift is of the magnitude expected if the ionization of a weak acid is important in catalysis, and the similar shape of the profile indicates that there is no drastic effect on the conformation of the various complexes in the mechanism.

The most striking feature of the kinetics of glucokinase in ^2H_2O is that the positive co-operativity with respect to glucose that is evident in ^1H_2O disappears in ^2H_2O, and is replaced by negative co-operativity (Fig. 2). Expressed somewhat differently, it appears that there is a small normal isotope effect of about 1.3 when the enzyme is saturated with both substrates, but a large inverse isotope effect of 3.5 at low glucose and high MgATP concentrations, i.e. the reaction is about 3.5 times faster in ^2H_2O than in ^1H_2O under these conditions. This inverse isotope effect is increased by lowering the concentration of MgATP, which suggests that it is an equilibrium isotope effect : this is because at vanishing MgATP concentrations glucose binding should be at equilibrium (regardless of whether the mnemonical model is correct) and thus any kinetic isotope effect on glucose binding should disappear.

A more detailed examination of the solvent isotope effects on the glucokinase reaction is given elsewhere (Pollard-Knight and Cornish-Bowden, 1984), including a "photon inventory analysis" using mixed solvents. Here we shall discuss how the major features of the results, as outlined above, are consistent with the mnemonical model for glucokinase. As the isotope effect is different in sign at high and low glucose concentrations, it seems clear that two separate effects are involved. We believe that the large inverse isotope effect may be the result of a stabilization of one form of free enzyme in ^2H_2O, or an increased rate of binding of glucose to the low-affinity form. For the former possibility, if the form of free enzyme released at the end of the catalytic cycle, postulated to be the less stable enzyme form, is stabilized in ^2H_2O the conformational change assumed to be responsible for the positive co-operativity in ^1H_2O might occur to a negligible extent in ^2H_2O. This would result in the loss of co-operativity and generate an inverse isotope effect, because the relaxation to a different enzyme form is assumed to be responsible for the fact that the reaction proceeds more slowly at low glucose and high MgATP concentrations in ^1H_2O than would be expected by extrapolating from the behavior at high concentrations of both substrates.

In the alternative explanation, the more stable form of free enzyme is postulated to bind glucose faster in ^2H_2O than in ^1H_2O, though the same free enzyme predominates at low glucose concentrations. The rate is then faster in ^2H_2O, generating an inverse isotope effect. The negative co-operativity arises from the assumption that the rate at which the more stable enzyme form binds glucose is greater than that of the less stable form in ^2H_2O. Thus the binding of glucose appears to become weaker with increasing glucose concentrations. In this explanation the behavior of glucokinase ^2H_2O resembles that described for wheat-germ hexokinase (Meunier et al., 1974).

The small normal isotope effect observed when the enzyme is saturated with substrates must be an effect on the catalytic reaction converting the ternary complex between enzyme, glucose and

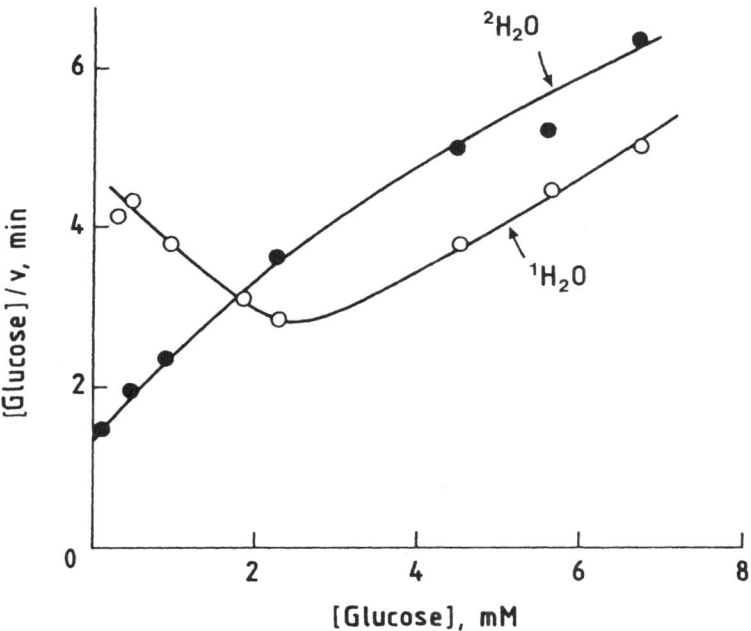

<u>Figure 2.</u> Negative co-operativity with respect to glucose in ^2H_2O. The characteristic upward curvature observed in ^1H_2O, indicating positive co-operativity, becomes inverted in ^2H_2O, indicating that there is now negative co-operativity. As the curves cross one another the reaction at very low glucose concentrations is about 3.5 times faster in ^2H_2O, i.e. there is a substantial inverse isotope effect. Measurements were done at 4.3 mM MgATP (near saturating) and the glucose concentrations indicated, at the pL optimum in each solvent, i.e. pL 8.0 in ^1H_2O and pL 8.4 in ^2H_2O, in 50 mM glycylglycine/KO^{1}H (or KO^{2}H) buffer containing 1 mM dithiothreitol and 0.1 M KCl.

MgATP into free enzyme and products. The relatively slow exchange between glucose and glucose 6-phosphate at chemical equilibrium (see above) suggests that release of glucose 6-phosphate from the enzyme may be rather slow, and thus probably rate-limiting when the enzyme is saturated with substrates. Thus it is likely that it is an effect of ^2H_2O on release of glucose 6-phosphate that is responsible for the normal isotope effect.

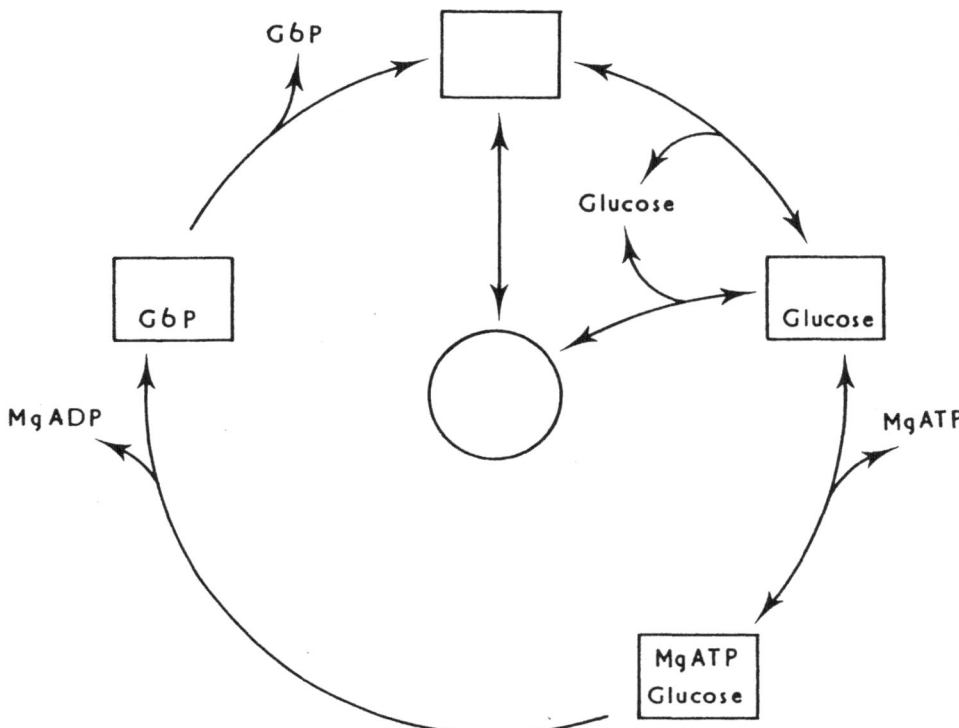

Figure 3. Mnemonical mechanism for glucokinase. The co-operativity in ^1H_2O is considered to arise from failure of the two forms of free enzyme to equilibrate with one another and with the enzyme-glucose complex when the MgATP concentration is high enough for it to react rapidly with the enzyme-glucose complex. It is assumed that the equilibrium constant between the two free enzyme forms favors the "circle" form and that this form binds glucose less strongly than the "square" form. At high MgATP concentrations the reaction proceeds predominantly through the "circle" form at low glucose concentrations, but increasingly through the "square" form as the glucose concentration increases ; thus the binding of glu-cose appears to become tighter as its concentration increases. If the primary effect of ^2H_2O is to alter the equilibrium constant between the free enzyme forms, so that the "circle" form is no longer favored, the pathway through this form would cease to be significant and the co-operativity would disappear. Alternatively, an increase in the rate constant for binding of glucose to the "circle" form in ^2H_2O would also result in loss of co-operativity and could also generate negative co-operativity.

These interpretations of the solvent isotope effects are illustrated in Fig. 3. Although they account for the gross features
of the results, a complete explanation is likely to be more complicated. In particular, the suggested effect on the equilibrium between the two forms of free enzyme can certainly account for the loss
of positive co-operativity, but it fails to explain why the co-operativity not only disappears but also becomes converted into negative co-operativity. To explain this it may well be necessary to
invoke binding of MgATP before glucose in the steady state : this
would agree with the results of isotope exchange at equilibrium,
but it is a complication that we shall not discuss in this chapter.

CONCLUSIONS

Although much has been learned about the glucokinase mechanism in recent years, there is still no direct evidence for the
existence of two forms of free enzyme that appear necessary to
explain the kinetics. A major difficulty in addressing this problem
by the well known techniques of protein chemistry is that the amount
of enzyme available for study is very small. The normal purification procedure for glucokinase (Holroyde et al., 1976) provides
a high yield and ample catalytic activity for many kinetic experiments, but a typical purification from the livers of 50 rats
provides only a few milligrams of protein. Thus it is not possible
to study glucose binding by equilibrium dialysis, etc. This problem
may perhaps be overcome by fluorescence measurements, but we have
only recently begun work on this. The initial results are promising
and we hope that in due course a full investigation by this technique will provide much fuller information about the binding of glucose to glucokinase.

ACKNOWLEDGEMENTS

The work described in this chapter was supported by a grant
from the Medical Research Council (U.K.) to A.C.-B., Dr. I.P.
Trayer and Professor D.G. Walker, and by a Science and Engineering
Research Council (U.K.) studentship to D. P.-K. We thank Dr. M.L.
Cárdenas for helpful discussion of the manuscript.

REFERENCES

Ainslie, C.R., Shill, J.P. and Neet, K.E. (1972) J. Biol. Chem.
247, 7088-7096.
Britton, H.G. (1966) Arch. Biochem. Biophys. 117, 167-183.
Cárdenas, M.L., Rabajille, E. and Niemeyer, H. (1978) Arch. Bio-
chem. Biophys. 190, 142-148.
Cárdenas, M.L., Rabajille, E. and Niemeyer, H. (1979) Arch. Biol.
Med. Exper. 12, 571- 580.
Connolly, B.A. and Trayer, I.P. (1979) Eur. J. Biochem. 99, 299-308.

Cornish-Bowden, A. and Gregoriou, M. (1981) Trends Biochem. Sci. 6, 149-150.

Ferdinand, W. (1966) Biochem. J. 98, 278-283.

Frieden, C. (1964) J. Biol. Chem. 239, 3522-3531.

Gregoriou, M., Trayer, I.P. and Cornish-Bowden, A. (1981) Biochemistry 20, 499-506.

Henderson, R.F. and Henderson, T.R. (1969) Arch. Biochem. Biophys. 129, 86-93.

Henderson, R.F., Henderson, T.R. and Woodfin, B.M. (1970) J. Biol. Chem. 245, 3733-3737.

Holroyde, M.J., Allen, M.B., Storer, A.C., Warsy, A.S., Chesher, J. M.E., Trayer, I.P., Cornish-Bowden, A. and Walker, D.G. (1976) Biochem. J. 153, 363-373.

Irias, J.J., Olmstead, M.R. and Utter, M.F. (1969) Biochemistry 8, 5136-5148.

Koshland, D.E. Jr., Némethy, G. and Filmer, D. (1966) Biochemistry 5, 365-385.

Meunier, J.C., Buc, J., Navarro, A. and Ricard, J. (1974) Eur. J. Biochem. 49, 209-223.

Monneuse-Doublet, M.E., Olomucki, A. and Buc, J. (1978) Eur. J. Biochem. 84, 441-448.

Monod, J., Wyman, J. and Changeux, J.P. (1965) J. Molec. Biol. 12, 88-118.

Neet, K.E., Ohning, G.V. and Woodruff, N.R. (1984) This volume.

Niemeyer, H., Cárdenas, M.L., Rabajille, E., Ureta, T., Clark-Turri, L. and Peñaranda, J. (1975) Enzyme 20, 321-333.

Olavarría, J.M. (1973) An. Acad. Cienc. Exact., Fis. Nat. (Buenos Aires) 25, 129-165.

Olavarría, J.M. (1980) Revta. Inst. Cibernetica Soc. Cient. Argent. 5, 88-102.

Olavarría, J.M., Cárdenas, M.L. and Niemeyer, H. (1982) Arch. Biol. Med. Exper. 15, 365-369.

Panagou, D., Orr, M.D., Dunstone, J.R. and Blakley, R.L. (1972) Biochemistry 11, 2378-2388.

Pollard-Knight, D. and Cornish-Bowden, A. (1982) Mol. Cell. Biochem. 44, 71-80.

Pollard-Knight, D. and Cornish-Bowden, A. (1984) Eur. J. Biochem., in press.

Rabin, B.R. (1967) Biochem. J. 102, 22C-23C.

Ricard, J., Meunier, J.C. and Buc, J. (1974) Eur. J. Biochem. 49, 195-208.

Shill, J.P. and Neet, K.E. (1975) J. Biol. Chem. 250, 2259-2268.

Smith, S. (1971) Biochim. Biophys. Acta 251, 477-481.

Storer, A.C. and Cornish-Bowden, A. (1976) Biochem. J. 159, 7-14.

Storer, A.C. and Cornish-Bowden, A. (1977) Biochem. J. 165, 61-69.

Taylor, K.B., Cook, P.F. and Cleland, W.W. (1983) Eur. J. Biochem. 134, 571-574.

Ureta, T. (1982) Comp. Biochem. Physiol. 71B, 549-556.

Walker, E.J., Ralston, G.B. and Darvey, I.G. (1975) Biochem. J. 147, 425-433.

Walker, E.J., Ralston, G.B. and Darvey, I.G. (1976) Biochem. J. 153, 329–337.
Wedler, F.C. and Boyer, P.D. (1972) J. Biol. Chem. 247, 984–992.

SECTION II – DYNAMICS OF SUBUNIT INTER-
ACTIONS IN MULTIMERIC ENZYMES AND THE
CONTROL OF CATALYSIS

SUBUNIT INTERACTIONS AND ENZYME KINETICS : THE LINKAGE APPROACH

E.P. Whitehead

Biology Department, Directorate-General of Scientific and
Technological Research, Commission of the European
Communities and Instituto di Chimica Biologica, Facoltà di
Medicina, Università di Roma, P.le A. Moro, Roma 00185

*Several advantages of basing analyses of steady-state kinetics
of allosteric enzymes on linkage principles are pointed out; among
them the ease and exactingness of tests that can be applied to the
data of limited accuracy normally available. These advantages are
illustrated by some analyses of the allosteric kinetics of deoxy-
cytidylate aminohydrolase (EC. 3.5.4.12).*

ADVANTAGES OF THE LINKAGE APPROACH

I might number the advantages of the linkage approach for
testing hypotheses about allosteric interactions as four :
SIMPLICITY, SECURITY, INSIGHT and SUGGESTIVENESS. In reality these
are all interconnected and are each aspect of one another.
1. Firstly there is the <u>simplicity</u> of the experiments, tests and
reasonings from experiment to interpretation
2. Security of the conclusions that can be drawn for linkage arguments, results from 1, from the relatively limited assumption that
go into the tests, and from the number and exactingness of the tests
that can be devised.
3. Then there is the physical and intuitive insight that the linkage approach brings, and
4. There follows from 3 its suggestiveness for devising experiments,
which brings us back again to 2.

With this preamble I might logically proceed to the principles
and philosophy of linkage, and explain what it is and why it works,

but I think that much of this, and why I claim the above advantages
will be more apparent and convincing if I proceed directly to a couple of practical illustrations.

EXAMPLES OF LINKAGE ANALYSIS

The two conformation hypothesis

Theory.
          Suppose we have an enzyme with allosteric sites for an activa-
tor A and an inhibitor I. Then the simplest model we have at present
to explain the activation and inhibition is that enzyme subunits
can exist in only two conformations, one of which binds only the
activator while the other binds only the inhibitor. How can we check
this model experimentally ? It could hardly be simpler.

          The kinetic information we would gather would usually be presen-
ted as a family of activation curves at constant inhibitor concentrations, or inhibition curves at constant activator concentrations
(Figs. 1A,B). The information in these two families is of course identical. They can both be considered representations of a surface in
three dimensions with the rectangular Cartesian axes $[A]$, $[I]$, and
velocity (v); the curves are then contours on the surface (as on
a relief map) those in Fig. 1A being contours of constant $[I]$
when the surface is viewed along the $[I]$ axis and similarly for
$[A]$ in Fig. 1B. Now it is of course not obvious just by looking at
these Figures whether they correspond to the above simple two-confor-
mational model or to some more complicated one (these are theoretical
curves calculated from a model).

          But it is much more obvious when the surface is viewed along the
v axis (Fig. 1C). In this case the co-ordinates of the plane Figure
are $[A]$ and $[I]$. Then the two-conformation model predicts
(Whitehead, 1973) that the contours (corresponding to constant v)
are straight lines. Furthermore all the straight lines meet in a
point, and the co-ordinates of this point are $(-K_A , -K_I)$, K_A and
K_I being the dissociation constants of A and I respectively for the
conformations that bind them.

          Fig. 1C contains the same information as Figs. 1A, B and can
be constructed as soon as one has obtained a family of activation or
inhibition curves experimentally. If one has obtained say seven such
curves, then for any given velocity one can obtain from experimental diagrams corresponding to Figs. A, B up to seven points
($[A],[I]$) corresponding to that velocity. The points shown in
Fig. 1C are those which would be obtained from the curves of Fig. 1A.

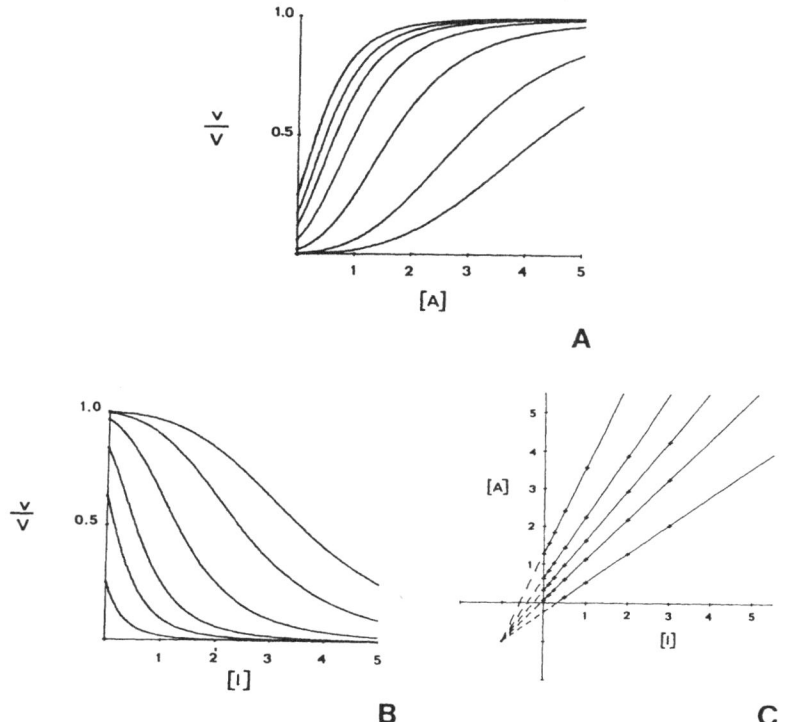

Fig. 1. The [A], [I], v/V surface viewed along the 3 axes.
Curves are calculated from a theoretical model with two conformations
and exclusive binding of activator and inhibitor to one and the
other conformation.
A. Family of activation curves at various fixed inhibitor concentra-
tions.
B. Family of inhibition curves at different activator concentrations.
C. The equal-velocity plot that would be obtained from Fig.1A. Each
line corresponds to a different fixed v/V (0.1, 0.3, 0.5, 0.7, 0.9).
The points marked on the lines are those which would be obtained from
Fig. 1A for these velocities.

Assumptions of the method

The above method which we may call the "constant velocity plot" has clearly the advantages of being simple and informative, but what assumptions does it depend on ? The answer is very few – another advantage to easy and general application. It does not depend on the nature of the subunit interactions – on whether these follow the Monod, Wyman, Changeux (1965) symmetric transition model, or a sequential pattern or, within the latter the interaction patterns postulated by Koshland et al. (1966), Cornish-Bowden and Koshland (1970), or an association-dissociation model (provided that the kinetic measurements are all done at the same enzyme concentration), or the "structural kinetic" models of Ricard et al. (1974); it does not depend on how the conformations interact with substrate – on whether one conformation or more binds and is catalytically active towards the substrate, indeed on any other substance present in the system at constant concentration; nor on knowing the number of effector sites (we assume that the number of identical sites for activator equals that for inhibitor though it does not matter whether the activator site is identical to the inhibitor site or not).

This, and why the method works may be clarified by a rough proof. Consider equilibria between the two conformations R and T of a single protomer at any $[A]$, $[I]$.

$$RA \;\rightleftharpoons\; R_o \;\rightleftharpoons\; T_o \;\rightleftharpoons\; TI$$
$$\quad K_A \qquad\qquad \emptyset \qquad\qquad K_I$$

(R_o, T_o are the concentrations of protomers unliganded by A and I in the R, T forms; R_{tot}, T_{tot} are the total concentrations of these forms). Velocity at constant substrate concentration depends only on the balance between the two conformations. As long as we do not alter this balance, velocity remains constant. Now,

(1) $$\frac{R_{tot}}{T_{tot}} = \frac{R_o + RA}{T_o + TI} = \frac{R_o(1 + [A]/K_A)}{T_o(1 + [I]/K_I)}$$

In an isolated protomer, obviously, if we vary $[A]$ and $[I]$ such as to keep the ratio of the bracketed factors above constant

(2) $$(1 + [A]/K_A) = C(1 + [I]/K_I)$$

where C is a constant, R_{tot}/T_{tot} will remain constant. But this is
also true when there are interactions between subunits. If there
is no tendency to vary the R/T balance of any individual subunit, then
no effect is transmitted to any other subunit; there is no alteration
in the total free energy of intersubunit interactions. Thus the existence of these interactions does not alter the fact that(2) corresponds to constant R_{tot}.

To put it another way, you will find that in all the different
models of intersubunit interactions, at constant concentrations
of other effectors

$$(3) \qquad v = f(\bar{R}) = F \left(\frac{1 + [A] / K_A}{1 + [I] / K_I} \right)$$

so that constant v corresponds to equation (2).
(For anyone unsatisfied by these arguments a more formal one is
given below).

This demonstration shows that the result is independent of all
the assumptions mentioned at the start of this subsection, since at
no point was it necessary to consider them. The really important
assumption we employ is that the effector interactions with the
enzyme can be treated as quasi-equilibria. Even this is less of a
limitation than it may seem, firstly in the comparative sense that
practically all other experimental steady-state kinetic analyses have
made it, and secondly only equations formally identical to quasi-
equilibrium equations are required. Such equations, without these
representing actual quasi-equilibrium, arise in sufficiently ordered
mechanisms (Dalziel, 1968; Whitehead, 1976) and in the "structural
kinetic" mechanisms of Ricard et al (1974), (Whitehead, 1976).
We would all be at sea if we had to deal with the true steady-state
equations of allosteric mechanisms; it seems best to see whether
the quasi-equilibrium assumption that simplifies these drastically
gives convincing results with predictive power.

Equ. 2 is my answer to the often heard view that steady-state
kinetics cannot contribute greatly to understanding allosteric
interactions : the equations are so complicated, involve so many
constants ... This simple equation shows that what matters is not the
number of constants but the form in which they enter into the equations. All of the constants except K_A and K_I, and the functions of
the concentrations of substrate and other effectors are tidied into
C, about which we needn't worry in isolating K_A, K_I and the typical
two-conformation pattern.

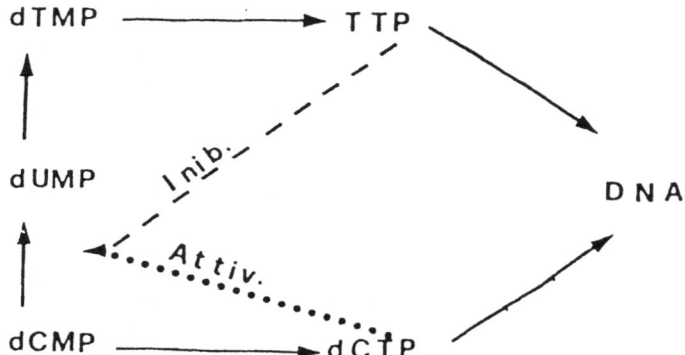

Fig.2. Partial metabolic scheme of pyrimidine nucleotide interconversions showing how the allosteric activation and inhibition by
triphosphates regulates their balance for DNA synthesis.

I believe in isolating the constants and functions that are easiest to study from the rest. A moment's thought shows that those which are easiest to study must also be those which most influence the overall behaviour of a system.

Application

We are at present applying this method to the dCMP aminohydrolase purified from donkey spleen. The allosteric activator is dCTP and the inhibitor dTTP. A biological purpose of these effects in regulating the pyrimidine deoxynucleoside triphosphate ratios is obvious from the metabolic scheme of Fig. 2. Less obvious may be that correct ratios of dNTP concentrations are important to faithful replication of DNA. When the ratios are different from those of the DNA to be synthesized, misincorporation by DNA polymerase giving incorrect base-pairings increases (Weinberg et al, 1981; Fehrsht, 1979). The explanation of this fact, by the way, involves an elegant piece of enzyme kinetics (Fehrsht, 1979; Fehrsht et al, 1982, 1983). It has been shown that in some human mutant cells lacking dCMP aminohydrolase, due to the loss of this allosteric control of dPyrNTP ratios, the spontaneous mutation rate increases about 10 fold (Weinberg et al, 1981).

We have applied the constant-velocity plot to this system, and found that in our conditions the plots are linear and converge closely towards a single point (unpublished work).

However a single result of this kind does not exploit the real power of the method. The pattern of Fig. 1C is independent of the concentrations of substrate and other effectors. Therefore if we change the substrate concentration the curves of Fig. 1 will change in detail, but the linearity of the constant-velocity contours should be unchanged and they should still converge on the same point as before. Likewise if we add new effectors provided that these do not induce new conformations. Thus we can do very may different sets of experiments, and the requirement that we obtain the same K_A, K_I from each set give us a very exacting test of the model (point 2 of the Introduction).

Such experiments are under way and the results will be published elsewhere. We have done rather more with this idea with another problem - that of the competition for substrate sites. The test of the two conformation hypothesis yields only two constants which lends both the analysis and the constant credibility; in the analysis to be given now only one physical constant is involved.

Tests of a model of competition for substrate sites
Theory

In a system with co-operative kinetics we clearly cannot apply classical competititve inhibition analysis. But we can still test a model of interactions between two inhibitors for the same site (Whitehead, 1970).

Suppose we want to test the hypothesis that a substrate, S, and a competitor, I, are bound exclusively to one and the same conformation and at the same (set of identical) sites. What does this model predict ?

Firstly it predicts that when either substrate or competitor are saturating all the enzyme is in the substrate-binding conformation, and we have classical competition, with the equation

$$(4) \qquad \frac{v}{V} = \frac{[S]}{[S] + a\,[I]}$$

Note that we cannot obtain the classical constants K_m and K_i because we are always at saturation, however we can obtain their ratio $a = K_m/K_i$.

In experiments at saturating concentrations with the substrate dCMP and the competitor dAMP we have found that Eq. 2 is very satisfactorily obeyed (see Fig. 3). The constant a can be obtained by conventional methods, and is subsequently used in analyses of behaviour at low concentrations (see below). On the other hand in similar conditions competition by dUMP (a product of the enzymatic reaction) does not at all obey this equation – both substrate saturation and product inhibition are co-operative. Thus the model survives for dAMP-dCMP interaction, and must be rejected for dUMP-dCMP interaction (there is reason to think that dUMP is in fact bound to the dCMP conformation, but also to other conformation(s)).

For a competitor like dAMP in this sytem, how can we analyse behaviour at lower concentrations where the effects of co-operativity are felt in the kinetics ?

Let $\bar{S}$, the saturation of substrate sites in the absence of the competitor be described by the function

$$(5) \qquad \bar{S}_{[I] = 0} = f([S])$$

If substrate is bound to only one conformation then (at least according to what most of us have always thought) fractional velocity must equal substrate saturation

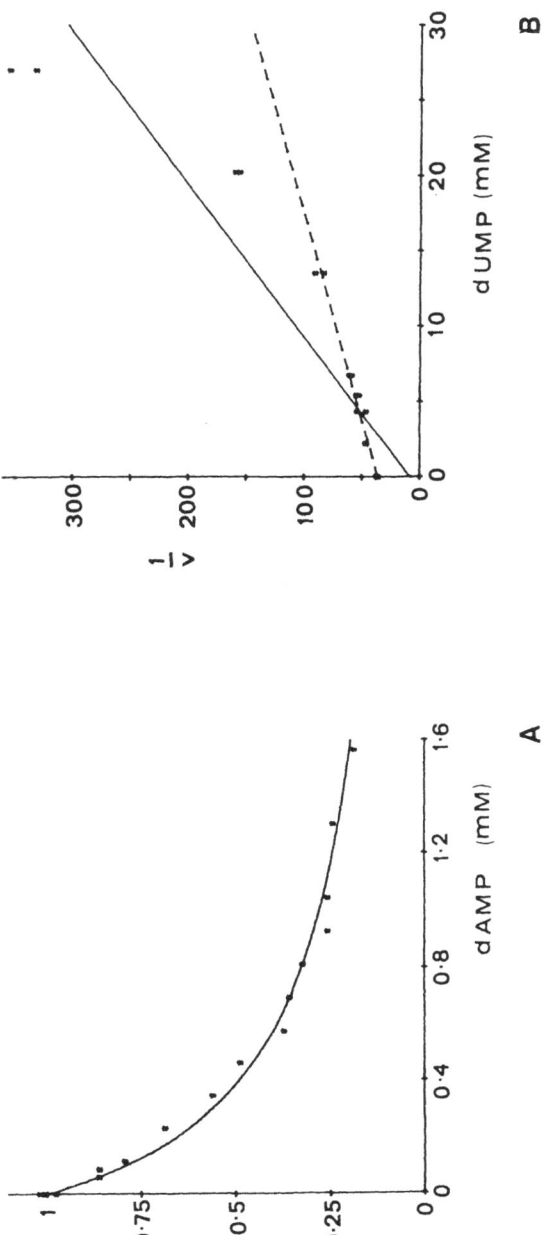

Fig. 3. Inhibition of dCMP aminohydrolase by substrate competitors at saturating substrate concentrations.

A . Typical fit of dAMP inhibition at saturating dCMP concentrations to Equ.2. The curve corresponds to Equ. 2 with values of a (here a = 5.47) and V obtained by fit of Dixon plot to data. dCMP concentration in this experiment was 2.14 mM.

Variance, $\sum (v_{th}/V - v_{exp}/V)^2/(N - 1)$ where v_{th} is the value by Equ. 2, v_{exp} the experimental estimate at the same dCMP, dAMP concentrations and N the number of experimental points (here N=17) was 0.87×10^{-3}.

B. Dixon plot of typical dUMP inhibition data at saturating dCMP (here 6.32 mM) showing impossibility of good fit to Equ. 2. The continuous line is a least-squares fit to all the data ; the broken line is a fit to the points up to 14 mM dUMP (corresponding to about V/2). The good linear fit down to about half-maximal velocity shows that the inhibition in this range is nearly non co-operative, however it becomes positively co-operative at higher [dCMP]/[dUMP] ratios. The velocity units are Δo.d. :min./mm lightpath at 290 nm.

(6)
$$\frac{v}{V} = \bar{S}$$

so let us write

(7)
$$\frac{v_{[I] = 0}}{V} = f([S])$$

It is fairly obvious that the binding of I in the absence of S
depends on the same function

(8)
$$\bar{I}_{[S] = 0} = f(a\ [I]\ )$$

(this equation also expresses that the co-operativity of substrate
binding and that of competitor binding, each in the absence of the
other is identical at the same saturation) and that the total binding when both are present together is given by

(9)
$$\bar{S} + \bar{I} = f([S] + a\ [I]\ )$$

The ratio of the binding of S and I to the whole enzyme is the
same as their ratio of binding to the single conformation that
binds them so

(10)
$$\frac{\bar{S}}{\bar{I}} = \frac{[S]}{a\ [I]}$$

v/V (and not, for example the binding of I) is what we can observe
in enzyme kinetics so we combine Eqs. 6, 9 and 10 to obtain :

(11)
$$\frac{v}{V} = \frac{[S]}{[S] + a\ [I]}\ f([S] + a[I]\ )$$

The models of Ricard et al (1974) do not agree with Equ.6 however
it can be shown that Equ. 11 would remain valid even with these
models (see Neet et al, 1984; Cornish-Bowden et al, 1984).

How can we use Equ. 11 for prediction? We have already seen that
we can obtain the constant a from experiments at high concentration.
(Note that for high enough values of the argument f = 1, and Equ. 11
reduces to Equ.4). But what about the function f ? This depends on
the model of subunit interactions, the constants, the number of sites,
etc. none of which has yet been defined. The answer is that f is suf-
ficently defined for our purposes by Equ. 7 : once we have obtained a
substrate saturation curve in the absence of competitor f is defined
empirically and this is all we need. Once we have this curve we can
for any value of ([S] + a[I]) determine f([S] + a[I]) from it, and
hence by Equ. 11 predict v/V for any pair of values ([S],[I]). This
brings out that linkage analysis consists essentially of comparing
curves with each other (as contrasted with the conventional model-
building approach which attempts to compare curves with the absolute
forms of algebraic equations).

Application

Fig. 4A shows that for the effects of dAMP on dCMP aminohydrolase , the predictions of Equ. 9 and experimental measurements are
in reasonable accord (Mastrantonio et al. 1983). (Note that, as
expected on this model, the competitor at low $[S]$, $[I]$, is an
activator, since as well as inducing the substrate binding conformation in its own subunit, the competitor through intersubunit co-
operativity also does so in other subunits making more sites available for substrate. At higher concentrations, the effect of competition for sites predominates and the competitor inhibits).

Here I can return to the idea at the end of the last section :
that we can make a test more exacting by adding other effectors to
the mixture. If these induce no new substrate or competitor-binding
conformations, then Equ. 11 still remains true, and the value of a
is unaltered. The only thing that changes is f which has to be rede-
termined in the new conditions. Beyond this empirical redetermina-
tion of the $v_{I=0}$, $[S]$ curve in the new conditions we need know nothing else for our purposes about the mechanism and type of interaction of the new effectors with the enzyme – we do not even need to
know whether they are bound at the substrate sites or elsewhere.

We have investigated the dCMP-dAMP interactions in the presence
of constant concentrations of dUMP (which is bound at substrate
sites) and the allosteric inhibitor dTTP and activator dCTP
(Mastrantonio et al, 1983). These completely change f : as well as
changing the concentration ranges in which the experiments are
done they change the co-operativity. In the absence of the effectors
the Hill coefficient of dCMP is in the range 2.6-2.9.dUMP reduces
substrate co-operativity at low dCMP:dUMP ratios, and gives almost
no co-operativity at the higher ratios (corresponding roughly to
the top half of the v, $[S]$ curve). dTTP increases substrate co-

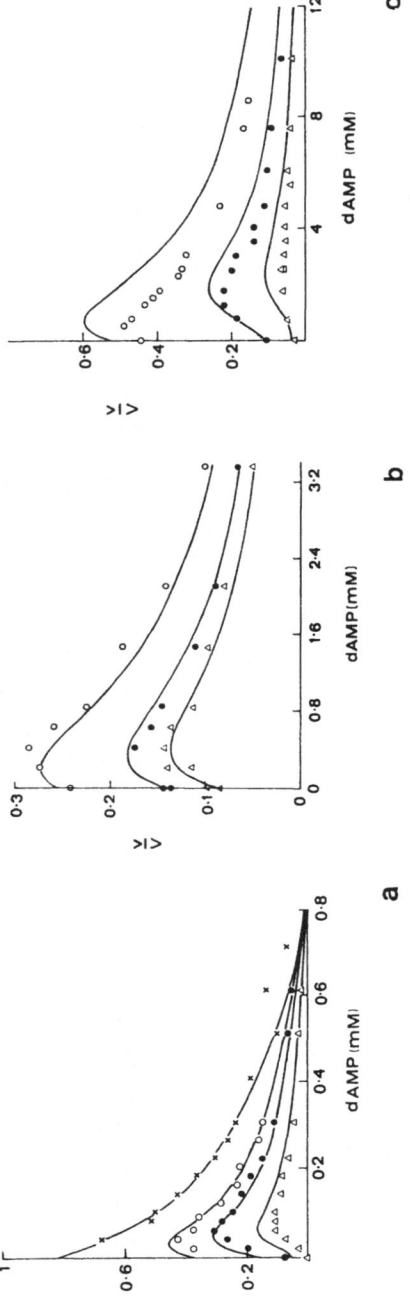

Fig. 4. Effects of dAMP concentration on rate of deamination of dCMP by dCMP aminohydrolase: comparison of predictions with experiments.

A. Effects in the absence of other effectors and at the following concentrations of dCMP : (Δ) 0.056 mM; (●) 0.108 mM ; (o) 0.158 mM; (X) 0.306 mM.

B. Effects in presence of 2.4 mM dUMP and the following concentrations of dCMP : (Δ)0.544 mM; (●) 0.726 mM; (o) 1.09 mM.

C. Effects in presence of 8 μM dTTP and the following concentrations of dCMP : (Δ) 1.14 mM, (●) 2.66 mM; (o) 6.08 mM. The curves are those predicted by Equ. 9 with the constant a and the function f determined as described in the text.

operativity (h is about 4) while dCTP abolishes it. Figs 4B , C
show that we still get adequate agreement between the predictions
of Equ.9 and experiment in the presence of dUMP, and of dTTP. (The
agreement in these Figures are not perfect, but experimental uncer-
tainties are such that we cannot say the deviations are significant.
Far more significant than any deviations is the first-order agreement
between experiments and the predictions of an equation with no free
parameters in four widely differing conditions, using the constant
a determined in the absence of the effectors). The kinetics in the
presence of dCTP not shown here (Mastrantonio et al, 1983) are a
special case, and indicate that dCTP is simply bound exclusively to
the same conformation as dCMP and dAMP. In its presence therefore
the kinetics reduce to classical competitive inhibition.

Thus we can be satisfied that dCMP and dAMP are bound to the
same conformation of the enzyme, both in the absence and presence of
several other effectors.

The philosophy of linkage

I hope the above examples suffice to convince of the advantages
of the linkage approach mentioned in Section 1. I might add that it
does not require exceptionally accurate rate measurements, nor complicated mathematical and numerical treatments, nonlinear curve-
fittings, etc... that make it difficult to trace critically how well
the evidence justifies some hypothesis.

Let us now step back and consider why the approach works and
what is behind it. Although I presented particular arguments for
the equations and methods used, both of them and anything else
one does with this approach stem from a single point of view,
indeed from a single equation, the famous "linkage equation"
(Wyman, 1964).

$$(12) \qquad \left(\frac{\partial X}{\partial Y} \right)_y = \left(\frac{\partial \ln y}{\partial \ln x} \right)_X$$

where X, Y are the amounts of any two ligands bound to a macromolecule, x and y their solution concentrations.

Apart from the fact that this is a very general equation that
all equilibrium binding to macromolecules is forced to obey whatever
the details, what is its importance for us ? We can think of this as
springing from the nature of our measurements and our objectives.

In a steady-state kinetics we only have one experimental output,
v/V which we can often assume to be equal to the fractional satura-

tion by substrate and is closely related to it. If x is substrate
and y is not, then we can measure X but not Y. Even in binding studies it is often much easier to measure the binding of one ligand
than another. So binding of for example, calcium ions to a respiratory protein will often be studied indirectly via the effect of solution $[Ca^{++}]$ on the more easily measurable oxygen binding. If we could
measure Y as easily as X, analyses of interactions would be considera-
bly simpler and the equations used more transparent.

But if we cannot measure Y, the linkage equation gives us information about it. We can know from experiment the quantity on the right
hand side of the equation and this tells us $\partial X/\partial Y$, how one ligand
is displacing the other from the protein or assisting its binding,
even though we only measure binding of one of these ligands. More
than this, we can know how one ligand displaces another when we can
measure the binding of neither - using the Linkage Equation for three
ligands x, y, z (Wyman, 1964) :

$$(13) \qquad \left(\frac{\partial Y}{\partial Z}\right)_{Y,z} = \left(\frac{\partial \ln z}{\partial \ln y}\right)_{x,X}$$

This is essentially what we are doing in the constant-velocity test.
Equ. 2 is just an integrated form of Equ. 13 when we feed in the two-
conformation hypothesis. For dA^{*} equals dR_{tot} multiplied by the saturation of the R conformation, $[A]/(K_A +[A])$, and similarly for I
except that the change in the amount of the I-binding conformation
is $dT_{tot} = -dR_{tot}$, thus

$$(14) \qquad \left(\frac{\partial A}{\partial I}\right)_{[A],[I]} = \frac{[A]/(K_A + [A])}{[I]/(K_I + [I])}$$

so by Equ.13 it follows that

$$(15) \qquad \left(\frac{\partial \ln [I]}{\partial \ln [A]}\right)_{[S],v} = \frac{[A]/(K_A + [A])}{[I]/(K_A + [I])}$$

which on integration gives Equ. 2.

---

*We now let X, Y, Z, x, y, z represent S, A, I, $[S]$, $[A]$, $[I]$.

Likewise Equ. 11 is an integrated form of the linkage equation (Equ.12) when the assumptions of the model are fed in (Whitehead, 1973).

The other important point here is the nature of our objectives. Most useful enzyme kinetic work involves study of the concentration effects of at least two effectors, say substrate and inhibitor , two substrates and two products. What we find out from this study is something about interactions between two or more substances when bound to the enzyme . Now however useful this has been, most would agree that enzyme kinetics is not the discipline which in itself brings us to the heart of the secret of enzyme catalysis, it is not the essence of this problem. On the other hand the important biological phenomena of allosteric control of enzymic activity is precisely a question of the enzyme-mediated interaction of different substances bound to enzyme, and so kinetics is the essence of this problem. And the most important part of this is how ligands displace or assist each other – in other terms is $\partial X/\partial Y$. (Not the whole story I know because there are also v effects and there may be non-equilibrium effects as well as equilibrium ones). From this I hope it is clear why I believe that enzyme kinetics far from being finished as some people think, really starts to come into its own with allosteric enzymes, and why linkage analysis takes us a long way along this road.

APPLICATIONS

In the following simple concrete cases it will be very evident that what really interests us is $\partial X/\partial Y$ and not the more-or-less complicated equations of models which threaten to distract us from this.

A classical behaviour in allosteric enzymes is that the co-operativity of a given effector is increased by the presence of antagonistic effector (in exclusive binding systems). Why this happens (and why the pattern is different in the case of nonexclusive binding, and why there is no change of co-operativity in the case of induced fit) is best understood through the linkage equations (Whitehead, 1973). The co-operativity can only increase up to a limit, after which it remains constant as the curves are shifted to the right.

It was Monod, Wyman et Changeux (1965) who first drew attention to this phenomenon, and in fact dCMP aminohydrolase was their principal experimental illustration. Here at high enough allosteric inhibitor (dTTP) concentrations the curves as Hill plots of the substrate (dCMP) saturation and the activation (by dCTP) , can be approximately described as having Hill coefficients, h_S , h_A of 4 which remains constant as increasing dTTP shifts the curves rightwards. Likewise the Hill slopes, h_I , of the inhibition curves are a constant -4^*. Here I will only consider this phenomenon at the extremes.

Now under the following conditions : that v/V is a function of substrate binding only (not necessarily equal to substrate saturation) and that the inhibitor can inhibit totally (which is the case in our system) then it is easy to show that

(16)
$$\frac{\partial S}{\partial I} = \frac{h_S}{h_I}$$

for

(17)
$$\frac{\partial S}{\partial I} = -\frac{\partial \ln [I]}{\partial \ln [S]} = \frac{\partial \ln [v/(V-v)]/\partial \ln [S]}{\partial \ln [v/(V-v)]/\partial \ln [I]} = \frac{h_S}{h_I}$$

(strictly speaking at a point, but to the extent that linearisation of the Hill plots is acceptable, for the curves). Thus the ratio of the Hill coefficients under these (admitedly not completely general) conditions tells us immediatly how the effectors are displacing or aiding each other.

Thus for substrate and inhibitor in our system

(18)
$$\frac{h_S}{h_I} = \frac{4}{-4} = -1$$

so

(19)
$$\frac{\partial S}{\partial I} = -1$$

so the data is trying to tell us that the binding of one molecule of inhibitor releases one molecule of substrate from the enzyme.

---

[*]It is convenient to define h as $d \log [(v - v_0)/(v_\infty - v)]/d \log x$ for any effector x (a slight break with the usual convention). Hill coefficients of inhibitors are then negative.

A similar argument applies to activator and inhibitor, so from the behaviour summarized above

$$(20) \qquad \frac{\partial A}{\partial I} = -1$$

each molecule of inhibitor displaces a molecule of activator and of course in these conditions too

$$(21) \qquad \frac{\partial A}{\partial I} = 1$$

so the binding of a molecule of activator is accompanied by that of a molecule of substrate.

Of course the conformational interpretation of this is evident, at effector concentrations much higher than their dissociation constants towards their appropriate conformations the R–T transition in any subunit implies the dissociation of one molecule of substrate (or activator) and the binding of one molecule of inhibitor : the concentration of unliganded subunits in either form is negligible and the equilibria may be schematised thus

$$RS + I \rightleftharpoons TI + S$$

or

$$RSA + I \rightleftharpoons TI + S + A$$

Note that we reach this interpretation straightforwardly without going through the more or less complicated equations of any particular subunit interaction model.

This shows how linkage clarifies the underlying physics of a classical behaviour, but let us now turn to something less well known.

Further investigation revealed that the description I gave above is only approximate – it occurs in the substrate concentration range 0.5 – 10 mM. But we can increase the allosteric inhibitor concentration so as to have substrate saturation curves that approach molar concentration for $S_{1/2}$. We find that over the higher range substrate co-operativity further increases to reach a Hill coefficient of around 8. The Hill coefficient of the inhibitor on the other hand at these substrate concentrations remains –4. Equ. 16 tells us

(22)
$$\frac{\partial S}{\partial I} = \frac{8}{-4} = -2$$

i.e. that a molecule of inhibitor displaces two molecules of substrate. We must conclude that there are two sites (with different affinities) on the protomer for substrate. We also conclude that the new substrate site we have invoked is not a catalytic site, as V remains constant over the whole range of these experiments. It is also implied that substrate is only bound to the second site when the protomer is in a catalytically active conformation, and equilibria can be schematized

$$RS_2 + I \rightleftharpoons TI + 2S$$

Why should the enzyme have a catalytically inactive site with a probably not physiologically relevant affinity for substrate ? It is only a small inductive leap , considering the similarity of dCMP to the activator dCTP, to suggest that this may be the activator site binding the substrate less well (by a factor of at least 10^5) than it does the activator. Now Equ. 16 gives us clear prediction. dCTP will compete with dCMP for the activator sites and at sufficient concentration prevent it binding there at all, so that substrate will be bound at only one site on the protomer. Thus at the same substrate concentration at which previously a molecule dTTP displace two of dCMP, it should now displace only one, and the h_s and $-h_I$ should again become equal. We have found that this is exactly what happens , and that in the presence of dCTP

$$h_s \simeq -h_I \simeq 4$$

according to the scheme

$$RSA + I \rightleftharpoons TI + S + A$$

This binding of the substrate to the active conformation of the allosteric site prompts the speculation that for this and many other allosteric enzymes,the active site and allosteric sites may have had a primitive active site as a common ancestor.

CONCLUSIONS

I hope that the above theory and experimental examples themselves speak enough of the advantages of the Linkage approach to steady-state allosteric enzyme kinetics. It is, so far as quasi-equilibrium formalisms hold, the most general approach to this, and within that same limitation, extracts most of the information that is present in the kinetic data. Of course having applied it does not prevent (and actually helps) anyone extracting the remaining information, which is quite limited and concerns the mode of intersubunit interaction.

Open general questions are : how far will we able to go using quite simple models like those considered above ; and how well will the quasi-equilibrium assumption continue to serve us ? Neither of the potential limitations implicit in these questions should discourage us from applying the linkage approach more widely, because the result of this application will itself answer the first question, and probably contribute to answereing the second. My feeling from the ongoing work on dCMP aminohydrolase that I have summarized above is that simple models are working adequately so far, and linkage has given us useful insight into the enzyme's behaviour.

ACKNOWLEDGEMENTS

I should like to express my appreciation of the contributions to the experimental work summarized above of my collaborator of the University of Rome, Drssa Stefania Mastrantonio, and of Prof. Mosé Rossi, Dr Roberto Nucci, Dr. Carlo Raia, Sgr. Carlo Vaccaro and Sgr. Santo Sepe in Naples.

This work was partly supported by Contract No BIO-E-399-I Commission of the European Communities - University of Rome. This is Publication No 2061 of the Biology dept., C.E.E.

REFERENCES

Cornish-Bowden, A. and Koshland, D.E. (1970) J. Biol. Chem.245 , 6241-6250.
Cornish-Bowden, A., Gregoriou, M. and Pollard-Knight, D. (1984) This Volume.
Fehrsht, A.R. (1979) Proc. Natl. Acad. Sci. U.S.A. 76, 4946-4950.
Fehrsht, A.R. , Knill-Jones, J.W. and Tsui, W.C. (1982) J. Mol. Biol. 156, 37-51.
Fehrsht, A.R., Shi, J.P. and Tsui, W.C. (1983) J. Mol. Biol. 165, 655-667.
Koshland, D.E., Nemethy, G.,and Filmer, D. (1966) Biochemistry 5 , 365-385.
Mastrantonio, S., Nucci, R., Vaccaro, C,, Rossi, M. and Whitehead, E.P. (1983) Eur. J. Biochem. in press.
Monod, J., Wyman, J. and Changeux, J.P. (1965) J. Mol. Biol. 12, 88-118.

Neet, K.E., Ohning, G.V. and Woodruff, N.R. (1984) This Volume.
Ricard, J., Mouttet, C. and Nari, J. (1974) Eur. J. Biochem. 41,
479-497.
Weinberg, G., Ullman, B., Martin, D. (1981) Proc. Natl. Acad. Sci.
USA 78, 2447-2451.
Whitehead, E.P. (1970) Biochemistry 9 , 1440-1453.
Whitehead, E.P. (1973) Acta Biol. Med. Germ. 31 , 227-258.
Whitehead, E.P. (1976) Biochem. J. 159 , 449-456.
Whyman, J. (1964) Adv. Prot. Chem. 19, 223-286.

KINETICS OF OXIDATIVE DECARBOXYLASES

Keith Dalziel

Department of Biochemistry, University of Oxford
South Parks Road, Oxford, England

INTRODUCTION

A common approach to understanding enzymic catalysis is to
dissect the overall reaction into likely steps and study the kinetics
of each of them in isolation by fast reaction techniques. For dehydrogenases, a discreet step is the dissociation of reduced coenzyme from its compound with the enzyme or with an enzyme-substrate
complex. This release of enzyme-bound coenzyme is often accompanied
by a change of its absorption spectrum or, more commonly and useful-
ly, by a change of its fluorescence emission spectrum (Boyer and
Theorell, 1956). This change can be used to monitor the reaction in
a suitable fast reaction apparatus, or to estimate the dissociation
constant of the complex by equilibrium measurements.

In the simplest case, the dissociation reaction may be a first-
order process and its reverse a second-order reaction, and one
can decide, for example, whether or not the release of reduced coenzyme is the rate-limiting step in substrate oxidation. If an enzyme molecule can combine with more than one coenzyme molecule, the
kinetics may be more complicated because of subunit interactions.
Even without subunit interactions, the kinetics may be complicated
because conformational changes of the protein molecule occur at a
different rate to the combination or dissociation of the coenzymes,
possibly accompanied by a further change of fluorescence of the
bound coenzyme. Similarly, the combination of the enzyme-reduced
coenzyme complex with another reactant —say the reduced substrate—
may cause another conformational change and therefore, almost certainly, a further change of coenzyme fluorescence in the complex,
and this probe can again be used to monitor the conformational
change.

My aim in this paper is to outline some of the results my colleagues and I have got during the last few years from studies of this kind with two NADP-linked oxidative decarboxylases, namely the malic enzyme from pigeon liver (Reynolds et al, 1978; Dalziel et al, 1983), and isocitrate dehydrogenase from bovine heart mitochondria (Dalziel et al, 1978, Fatania et al, 1982a, 1982b). It turns out that the rate limiting steps in these analogous enzyme-catalyzed reactions are quite different, and that only the malic enzyme shows good evidence of subunit interactions, which manifest themselves as half-of-the-sites reactivity.

MALIC ENZYME

All the experiments described were made in 0.05M tri-ethanola-mine buffer, pH 7.0 and at 30°C. From equilibrium studies it appears that 4 NADPH (or NADP) molecules can bind independently and with equal affinities to the 4-subunit enzyme (Pry and Hsu_2 1980). On the other hand, malate and the essential metal ion Mn^{2+} show negatively co-operative binding in which two sites on the enzyme have much greater affinities than the remainder (Hsu et al, 1976; Pry and Hsu, 1980).

Kinetics of NADPH dissociation

The independent binding of 4 NADPH molecules is reflected in the kinetics of its release from its enzyme complex, which is a simple first-order reaction, as shown in Figure 1A. This is a pro-gress curve obtained after mixing about 10 μM enzyme-NADPH complex from one syringe of a stopped-flow apparatus with 1mM-NADP from the other syringe. The decrease of fluorescence of NADPH, as it disso-ciates from the enzyme and is replaced by NADP, was measured at wavelengths greater than 400 nm, with excitation at 340 nm. Usually 1000 digital voltage values corresponding to the changing fluorescence, F_t, were stored on computer disc and analyzed by non-linear regression to the function

$$(1) \qquad F_t = F_0(A_1 e^{-k_1 t} + A_2 e^{-k_2 t} + A_3)$$

A satisfactory fit was obtained to Figure 1A with $k_2 = 0$, as shown in Figure 1B, and a mean first order rate constant of $k_1 = 418 \pm 23s^{-1}$ was obtained from 4 such progress curves in this experiment. Closely similar values were obtained in further expe-riments with 0.5, 4.0 and 10 mM NADP, and with 1 mM 2-monophospho-adenosine-5'-diphosphoribose (ATPR) as displacer (Table 1), showing that dissociation of NADPH from its enzyme complex is the rate-limiting step in the displacement reaction. The mean value of $427s^{-1}$

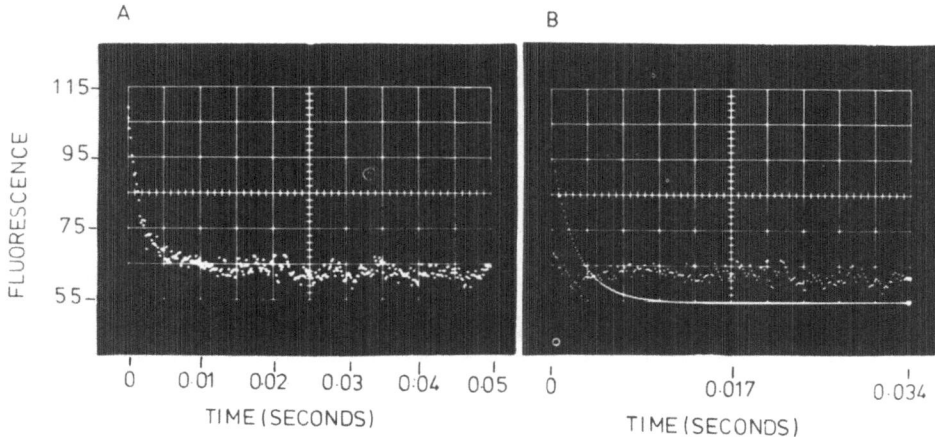

Fig. 1. A, The time course of the decrease of Fluorescence accompanying the displacement of NADPH from its complex with malic enzyme (10 μm) by NADP (1 mM). B. Simulated progress curve and residuals obtained by non-linear regression to a single exponential function. Reprinted by permission of Dalziel et al, 1983.

Table 1. Apparent first-order rate constants for the dissociation of NADPH from its complex with malate enzyme[a].

[enzyme] (μM)	[NADPH] (μM)	[NADP] mM	[ATRP] mM	k (s⁻¹)
8.75	8.35	0.5	--	436 ± 26
10.0	9.5	1.0	--	418 ± 23
10.0	9.5	4.0	--	431 ± 72
10.0	9.5	10.0	--	463 ± 82
10.0	9.7	-	1.0	385 ± 12
			Mean	427

[a]Progress curves for the reaction were obtained by recording the decrease of fluorescence in the stopped-flow apparatus after mixing the enzyme-NADPH compound with an excess of NADP or ATPR. The rate constant was estimated by non-linear regression of 4-8 progress curves obtained in each experiment to a single exponential function.

is therefore a good estimate of the rate constant for the dissociation of each coenzyme molecule from E_4-NADPH$_4$.

Figure 2 shows the results of a similar experiment in which
0.2 mM Mn^{2+} was also present. This concentration is sufficient to
saturate the two tight metal-binding sites (K_d = 8 µM) but not the
weak binding sites (K_d = 900 µM). It clearly slows down the NADPH
dissociation. Non-linear regression to a single exponential function
(Figure 2B) gives k_1 = 136s^{-1}, and from 5 such experiments in which
1 mM NADP or different concentrations of ATPR were used, a mean
value of 135 s^{-1} was obtained (Table 2). However, inspection of
the residuals in Figure 2B suggests that there is a faster process
occurring in the first few milliseconds, and Figure 2C shows the

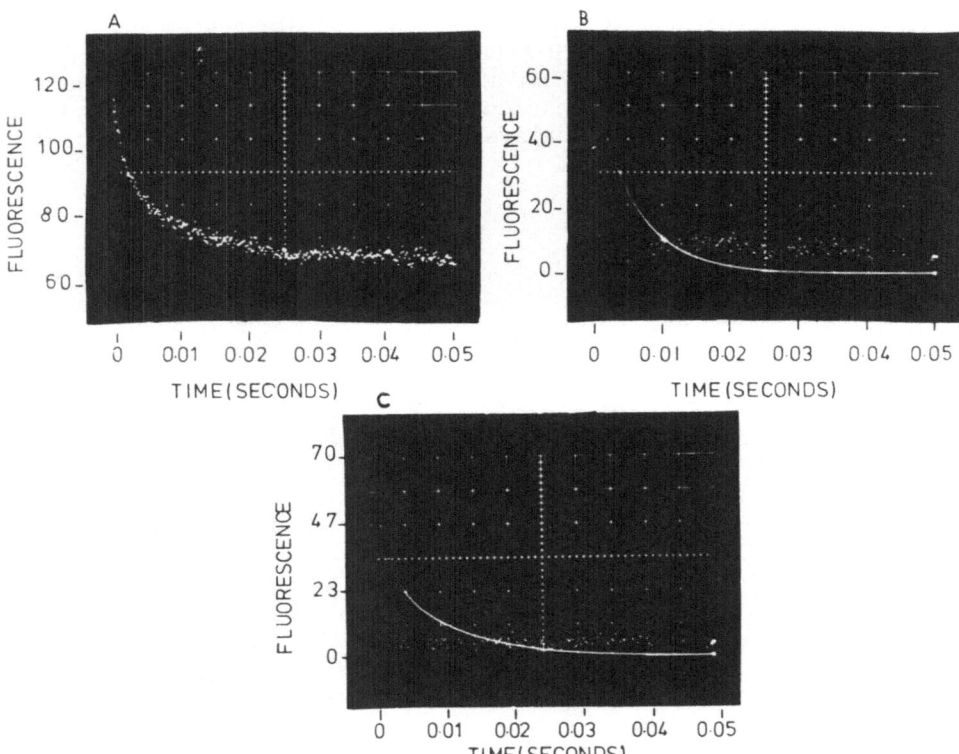

Fig. 2. A. Kinetics of the dissociation of NADPH from its enzyme
complex in the presence of 0.2 mM Mn^{2+}. B. Simulated curve and
residuals obtained by non-linear regression to a single exponential
function. C. Simulated progress curve and residuals obtained by
non-linear regression to equation (1). Reprinted by permission from
Dalziel et al, 1983.

Table 2. Apparent first-order rate constants for the dissociation of NADPH from its enzyme complex in presence of 0.2 mM Mn^{2+} [a].

[enzyme] (μM)	[NADPH] (μM)	Displacer	k (s^{-1})
10.0	11.0	1 mM NADP	130 ± 5
10.0	11.0	0.83 mM ATPR	130 ± 4
10.0	11.0	3.2 mM ATPR	127 ± 11
7.4	8.8	1.0 mM ATPR	105 ±
10.0	9.5	1.0 mM ATPR	182 ± 8
		Mean	135

[a] In each experiment the rate constant was estimated by fitting from 4 to 8 progress curves to a single exponential function by non-linear regression.

residuals from non-linear regression to the sum of 2 exponentials (equation 1), which fits the data better. A mean value of $96s^{-1}$ is obtained for the slower process in this way. This is 0.45 times the maximum turnover number for the tetrameric enzyme molecule in catalysis. A mean value of $800s^{-1}$ is obtained for the faster process, nearly 90% of which occurs in the dead-time of the stopped-flow apparatus of 2 milliseconds. The measured amplitude of the fast process is therefore variable (Table 3) but is about half the total amplitude of the displacement reaction. Its significance is uncertain, but we think it may represent dissociation of NADPH from $Mn2+$ - deficient subunits, which will certainly be present with this low Mn^{2+} concentration of 0.2 mM.

The cause of substrate inhibition

Inhibition of catalysis by malate concentrations greater than 0.3 mM (Hsu et al, 1976) was correlated with the formation of an abortive complex E - Mn^{2+}- NADPH-malate by Reynolds et al (1978). It has now been established that the formation of this complex is the cause of substrate inhibition by measurements of the rate of dissociation of NADPH from it. Figure 3A shows the kinetics of the displacement of NADPH from the enzyme by 1 mM ATPR in the presence of 0.2 mM $MnCl_2$ and 1 mM malate. More than 90% of the observed fluorescence decrease can be described by a first-order process with a rate constant of $2.72s^{-1}$. The results of a similar experiment with 0.3 mM malate (Figure 3B) are biphasic, and non linear regression to equation (1) indicates two first-order processes with $k_1 = 100s^{-1}$ and $k_2 = 2.69s^{-1}$. These values were obtained as means from 8 experiments with different malate concentrations, in which

Table 3. Interpretation of progress curves for the dissociation of NADPH in presence of 0.2 mM Mn^{2+} as the sum of two exponentials

A_1	A_2	k_1 (s^{-1})	k_2 (s^{-1})
406 ± 65	387 ± 20	846 ± 170	97 ± 5
381 ± 77	357 ± 46	804 ± 331	96 ± 9
345 ± 80	339 ± 16	684 ± 292	89 ± 11
406 ± 52	220 ± 28	645 ± 150	99 ± 17
			Mean 96

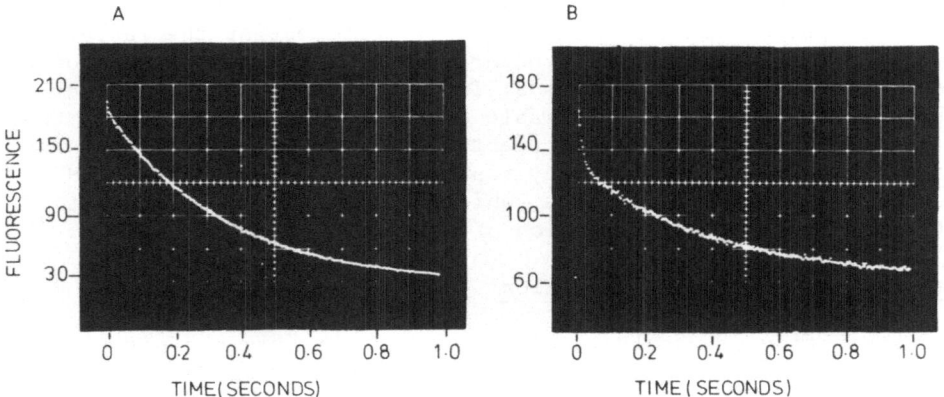

Fig.3. Progress curves for the dissociation of NADPH from its enzyme complex in the presence of 0.2 mM Mn^{2+} and (A) 1 mM malate (B) 0.3 mM malate. Reprinted by permission from Dalziel et al, 1983.

the amplitude of the fast phase decreased, and that of the slow phase increased, as the malate concentration was increased from 0.075 mM to 7 mM.

It was shown by equilibrium dialysis (Pry and Hsu, 1980) that in the presence of Mn^{2+} and NADPH, two molecules of malate bind tightly to the tetrameric enzyme, with K_d = 23 μM, and two more weakly, with K_d = 400 μM . We therefore conclude that the rate constant of $100s^{-1}$ should be assigned to NADPH dissociation from $E_4 - Mn_2^{2+}$-$NADPH_4$-$malate_2$, containing tightly-bound malate, and is the same as the rate constant of $96s^{-1}$ for NADPH dissociation from the malate-free complex $E-Mn_2^{2+}$-$NADPH_4$. The rate constant of $2.69s^{-1}$ is attributed to NADPH dissociation from $E_4-Mn_2^{2+}$-$NADPH_4$-$malate_4$, in which the weak sites for malate are also saturated. We conclude that combination of the weakly bound malate at the two metal-free subunits induces a conformation change in all 4 subunits that greatly decreases the rate of NADP dissociation and causes inhibition.

High Mn^{2+} concentrations decrease malate inhibition, but also decrease the fluorescence enhancement of NADPH in the enzyme complex so that it is not possible to study the kinetics of NADPH release with high Mn^{2+} concentrations by stopped-flow fluorescence measurements.

Time-course of the conformation change caused by malate

The early time-course of NADPH formation in the enzyme catalyzed oxidative decarboxylation of malate was studied by initial-rate measurements in the stopped-flow apparatus using absorbance measurements at 340 nm. A solution containing 1.14 μM enzyme active centres, 1 mM NADP and 0.2 mM or 0.06 mM $MnCl_2$ was put in one syringe of the apparatus, and one containing $MnCl_2$ and various concentrations of malate in the other. With malate concentrations in the reaction mixture of 0.3 mM or more there was a decrease of rate during the first half second (Figure 4A), whereas with smaller malate concentrations the progress curve was linear (Figure 4B). Plots of the reciprocals of the true initial rates against the reciprocal malate concentrations were linear, whereas using the steady state rates reached after 0.5s or more gave curved plots typical of substrate inhibition, in confirmation of earlier work.

The development of substrate inhibition during the first half-second may be interpreted in the following manner (cf. Dalziel et al, 1978). The rate of NADPH formation in the early stages of the reaction will be

(2) $$\frac{d[NADPH]}{dt} = [E_1]a + [E_2]b = [E_T](b + (a - b)\,e^{-kt})$$

where E_1 and E_2 are the concentrations of the normal and inhibited enzymes with specific activities a and b respectively, and k is the apparent first-order rate constant for the conversion of E_1 into E_2. By integration

(3) $$[NADPH]_t = [E_T]\left\{ bt + \frac{(a - b)}{k} - \left(\frac{a - b}{k}\right) e^{-kt} \right\}$$

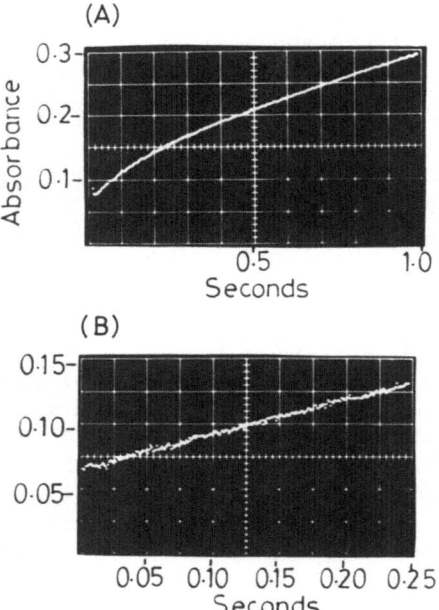

Fig.4. Progress curves for the oxidative decarboxylation of malate. The stopped-flow apparatus reactant concentrations were 0.57 M enzyme subunits recorded in 0.5 mM NADP, 0.2 mM Mn Cl_2 and A. 2.5 mM malate, B. 0.1 mM malate. Reprinted by permission from Dalziel et al, 1983.

When t is large, the exponential term will be negligible and the progress curve linear, as described by

(4) $$[NADPH]_t = E_T \, bt + \frac{(a - b)}{k}$$

If this linear portion is extrapolated back to smaller values for t, the difference between the extrapolated and observed values of the absorbance will be

(5) $$A_{ext} - A_{obs} = \left(\frac{a - b}{k}\right) e^{-kt}$$

or

(6) $$\log \Delta A = \log \frac{(a - b)}{k} - \frac{kt}{2.3}$$

A plot of log ΔA <u>versus</u> t from a progress curve obtained with 0.06 mM $MnCl_2$ and 5 mM malate is shown in Figure 5, and from the slope a value of $6.9 s^{-1}$ is obtained for k. With 0.06 mM and

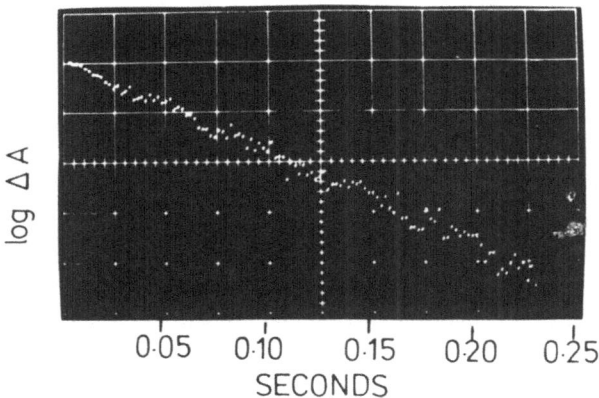

Fig. 5. A plot of log ΔOD <u>versus</u> time for a progress curve obtained with 0.06 mM $MnCl_2$, 5 mM malate, 0.57 mM enzyme subunits and 0.5 mM NADP. ΔOD is the difference between the absorbance obtained by extrapolation of the later linear part of the progress curve and the measured value. Reprinted by permission from Dalziel et al, 1983.

0.2 mM Mn^{2+} and malate concentrations varying 8-fold from 0.6 to
5.0 mM, similar plots were linear and k varied from 5.0 to 7.1,
which is not a significant variation. We conclude that this is
the rate constant for the conformation change which follows combination of malate with the enzyme-NADPH product complex, and results
in a slower release of NADPH from the enzyme.

The rate-limiting step and half-of-the-sites reactivity

With very large malate concentrations, the steady-state rate
of malate oxidation is 93% inhibited (Hsu and Pry, 1980). Kinetic
studies now show a parallel decrease in the dissociation rate of
NADPH from the enzyme from $100s^{-1}$ to $2.69s^{-1}$ as the malate concentration is increased. This is a strong evidence that NADPH dissociation is the rate-limiting step in the oxidative decarboxylation
reaction. The discrepancy between the measured rate constant for
NADPH dissociation from the complex $E_4 - Mn_2^2 - NADPH_4$ of $96s^{-1}$ and
the turnover number in catalysis of $53\pm3s^{-1}$ per subunit (Dalziel
et al, 1983) is most simply resolved by the assumption of half-of-the-
sites reactivity. This is in accord with earlier evidence, which
includes a maximum burst of enzyme bound NADPH equal to half the
sites (Reynolds et al, 1978), half-sites reactivity of bromopyru-
vate in affinity labelling, and half-of-the-sites binding of the
transition state analogue oxalate (Pry and Hsu, 1980; Hsu and
Pry, 1980, Vernon and Hsu, 1983). In this model , the rate-limiting
step is NADPH dissociation from $E_4 - Mn_2^{2+} - NADP_2 - NADPH_2$, and this
is regulated when the malate concentration is large and the Mn^{2+}
concentration small by conversion to the abortive complex and
slow NADPH dissociation.

ISOCITRATE DEHYDROGENASE

The experiments described were made in 0.05 M tri-ethanolamine
buffer, pH 7.0, or in 0.1 M phosphate buffer, pH 7.0, at 25°C.

Time course of a conformation change caused by isocitrate and Mg^{2+}

Progress curves for the oxidative decarboxylation of isocitrate in 0.05 M triethanolamine measured in the stopped-flow apparatus
are not linear if the enzyme is first mixed with excess NADP and
Mg^{2+} , and then with excess isocitrate and Mg^{2+}. The rate increases
more than two-fold during the first four seconds of the reaction,
and then remains constant (Figure 6). In 0.1 M phosphate, on the
other hand, such preincubation does not result in a non-linear
progress curve. If the enzyme is first mixed with isocitrate and
Mg^{2+}, and then with NADP, the progress curves in both buffers are
also quite linear from 2 milliseconds onwards. Pre-incubation with
NADPH alone in 0.05 M triethanolamine buffer also resulted in an
increase of the rate of isocitrate oxidative decarboxylation during

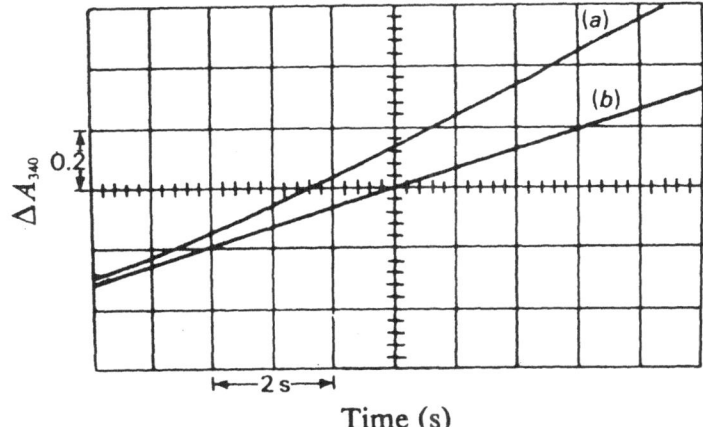

Fig. 6. Effects of pre-incubation of isocitrate dehydrogenase with
(a) 0.5 mM NADP and 5 mM MgCl$_2$ and (b) 0.9 mM–isocitrate and
5 mM MgCl$_2$ on the time-course of NADP formation in 0.05 M triethanolamine buffer, pH 7.0. Reprinted by permission from Dalziel et al,1978.

the first 4 seconds after mixing with NADP, Mg^{2+} and isocitrate
(Figure 7). However pre-incubation with NADPH, Mg^{2+} and isocitrate,
which form an abortive complex with the enzyme, did not induce the
lag phase.

    In Figure 8, plots of log(A$_{obs}$ – A$_{ext}$) versus t are shown for
3 experiments, in which the enzyme was pre-incubated with NADP,
with NADPH and NADP, and with NADP, Mg^{2+} and 2-oxoglutarate. In
accordance with the hypothesis already outlined for the malic enzyme, the linear plots are consistent with a first-order conformation
change of the enzyme, and the slopes give a value of 0.40±0.05s^{-1}
for the rate constant for the isomerization, which is presumed to
follow the combination of isocitrate and Mg^{2+} with either the enzy-
me-NADP or the enzyme-NADPH complex in catalysis.

Kinetics of NADPH dissociation

    The time course of the displacement of 6 µM NADPH from its

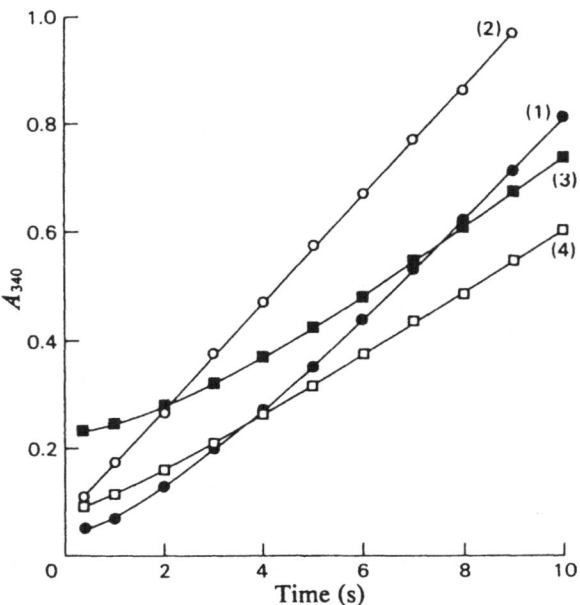

Fig.7. Effects on the progress curve in 0.05 M tri-ethanolamine buffer of pre-incubating the enzyme with NADP (●) with isocitrate and $MgCl_2$ (O), with NADPH (□), and with buffer alone. Reprinted by permission from Dalziel et al, 1978.

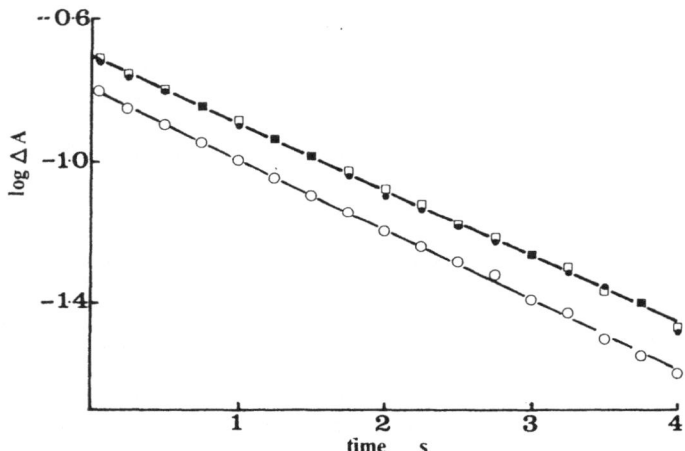

Fig.8. A plot of log ΔOD (defined in the legend to Figure 5) versus time for the isocitrate dehydrogenase reaction after pre-incubation of the enzyme with NADP(o), NADPH and NADP (□), and NADP, $MgCl_2$(●) and 2-oxoglutarate (o). Reprinted by permission from Dalziel et al, 1978.

complex with the enzyme by 4 mM NADP, in 0.05 M tri-ethanolamine
and 5 mM $MgCl_2$, is shown in Figure 9A by the decreasing coenzyme
fluorescence. The first-order plot of the data in Figure 9B shows
that the reaction can be described in terms of 2 first-order processes, with rate constants of $0.6s^{-1}$ and $5s^{-1}$. The amplitude of
the faster process is about one-fifth of the total amplitude of the
reaction. Since the turnover number per subunit for the oxidative
decarboxylation reaction, from initial rate measurements in the
steady state, is $18s^{-1}$, it appears that the dissociation of NADPH
from the complex $E-Mg^{2+}$ - NADPH is too slow for this to be a step
in catalytic cycling in this buffer.

The results of a similar experiment in 0.1M phosphate are
shown in Figure 10. The time-course of the fluorescence change
again indicates two first-order processes, with rate constants of
$0.7s^{-1}$ and $16s^{-1}$, but in this case the amplitude of the faster
process is two-thirds of the total amplitude of the reaction. These
results were confirmed in several experiments with different concentrations of the displacing reagent NADP (Fatania et al, 1982a).

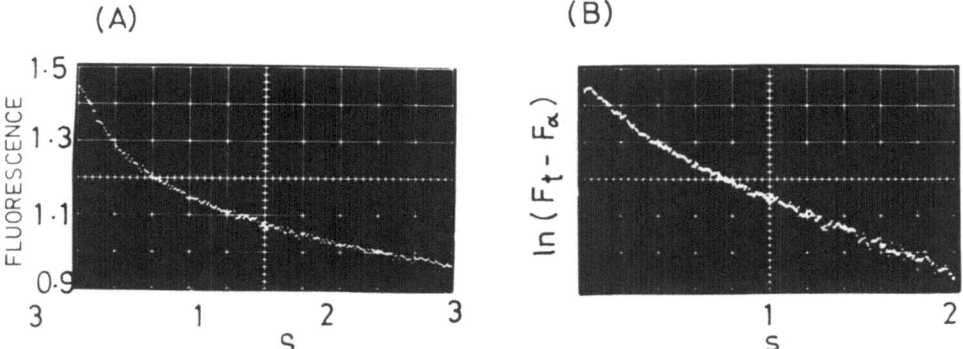

Fig. 9. A. Progress curve for the displacement of NADPH from its
complex with isocitrate dehydrogenase in 0.05M tri-ethanolamine
buffer pH 7.0 containing 5 mM $MgCl_2$ by 4 mM NADPH. B. A first-order
plot of the results,

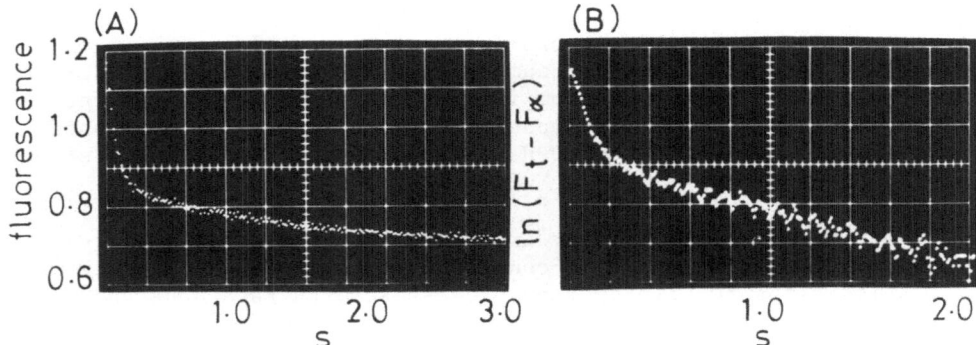

Fig.10. A. Progress curve and B. first-order plot for the displacement of NADPH from its complex with isocitrate dehydrogenase in 0.1 M
phosphate buffer, pH 7.0, containing 5 mM MgCl$_2$. Reprinted by permission from Fatania et al, 1982a.

The simplest interpretation of the complexity of the NADPH
dissociation reaction is that the two subunits in the enzyme molecule have different rate constants, either intrinsically or because
of subunit interactions. It is not possible, however, to explain
the substantial differences between the amplitudes of the fast and
slow phases in terms of the integrated rate equation for this
model (Fatania et al, 1982a).

An alternative hypothesis is that the enzyme-NADPH complex
exists in 2 conformations in equilibrium, one of which dissociates
NADPH much more rapidly than the other. Thus,

$$\text{E}^{*}.\ \text{NADPH} \underset{\longleftarrow}{\xrightarrow{0.65\text{s}^{-1}}} \text{E.NADPH} \xrightarrow{16\ \text{or}\ 6\text{s}^{-1}} \text{E+NADPH}$$

We have shown that the kinetic data obtained both in phosphate and in tri-ethanolamine buffer are satisfactorily accounted
for by the integrated rate equation for this model (Fatania et al,
1982a). The model also provides an explanation for the lag phase

in the oxidative decarboxylation reaction which results from pre-
incubation of the enzyme with NADPH, or NADP, in tri-ethanolamine
buffer, but not in phosphate buffer. The conformation $E^{\bigstar}$.NADPH
accounts for 80% of the enzyme in tri-ethanolamine buffer, according
to the kinetics of NADPH dissociation, whereas in phosphate buffer
only 30 % of the enzyme is in this form. The lag phase in catalysis
therefore represents the transition from $E^{\bigstar}$.NADPH which follows combination of isocitrate with the enzyme.

The rate-limiting step in catalysis

Dissociation of NADPH from $E-Mg^{2+}$-NADPH in tri-ethanolamine
buffer is too slow for this reaction to be a step in catalysis.
However, the abortive complex $E-Mg^{2+}$ -NADPH-isocitrate is known to
have a dissociation constant for isocitrate of only 4 μM (Reynolds
et al, 1978b) while K_M for isocitrate is 10 μM.The formation of
considerable amounts of the abortive complex in steady-state catalysis is therefore to be expected. Moreover, the rate of dissociation of NADPH from the abortive complex is very large, about 300-
$500s^{-1}$. This is shown in Figure 11, where the time course of NADPH
dissociation which follows mixing of E.NADPH with excess isocitrate and Mg^{2+} in the stopped-flow apparatus is shown, again by the
decrease of the coenzyme fluorescence as it is set free. This
experiment is based upon the fact that 90% of the enzyme and NADPH

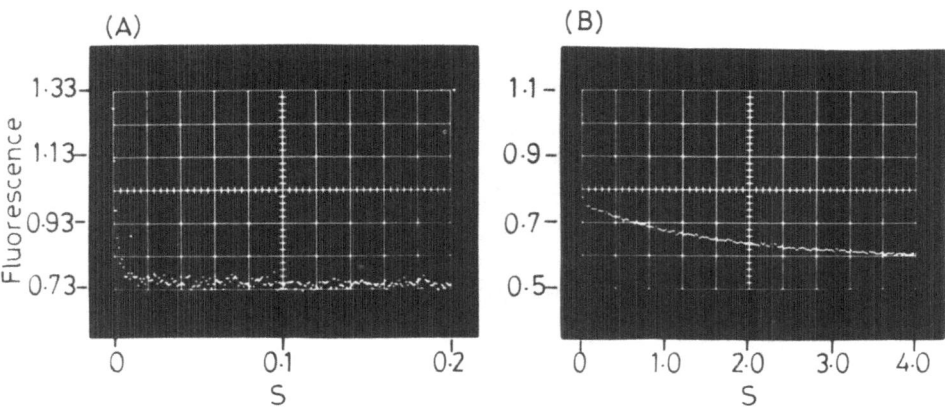

Fig. 11. The kinetics of dissociation of NADPH from the abortive
complex with enzyme, Mg^{2+} and isocitrate 0.1M phosphate buffer,
pH 7.0. The sum of the fluorescence of the reactants before mixing
was 1.33. Reprinted by permission from Fatania et al, 1982a.

will be complexed initially, while not more than 20% will remain
bound after the addition of isocitrate and Mg^{2+} because of the
large dissociation constant (20 µM) for NADPH from the abortive
complex (Reynolds et al, 1978b).

We conclude that the product complex E.Mg NADPH is converted
into the abortive complex in catalytic cycling in both buffers, and
probably _in vivo_ also, and that NADPH dissociates rapidly from
this complex. Therefore an earlier step must be slow enough to limit the maximum turnover to $18s^{-1}$. We have confirmed the occurence
of this slower step by single turnover experiments (Fatania et al,
1982b) but have not yet established its chemical nature.

CONCLUDING COMMENTS

Evidently the rate-limiting steps in the reactions catalyzed
by the malic enzyme and isocitrate dehydrogenase are quite different. Moreover, whilst both enzymes form abortive complexes
containing reduced coenzyme and reduced substrate, which undergo
relatively slow conformational changes, the abortive complex disso-
ciates NADPH more slowly than does $E-Mn^{2+}$-NADPH in the case of
malic enzyme, but more quickly than $E-Mg^{2+}$-NADPH in the case of
isocitrate dehydrogenase. It is unlikely that the conformation
changes of isocitrate dehydrogenase, induced by coenzyme binding and
reversed by isocitrate binding, have any relevance to the behaviour
of the enzyme in the mitochondrion where most of the enzyme will
probably be complexed with isocitrate. It is also difficult to
visualise a detailed role for substrate inhibition or half-of-the-
sites reactivity for the malic enzyme in the living cell. It is
expected that the combination of any ligand with a mobile protein
molecule will result in a change of conformation, indicated in the
present work by a changed yield of NADPH fluorescence. But not all
such changes will be relevant to catalysis.

REFERENCES

Boyer, P.D and Theorell, H. (1956) Acta Chem. Scand. 10 , 447-
450.
Dalziel, K., McFerran, N., Matthews, B. and Reynolds, C.H. (1978)
Biochem.J. 171 ,743-750.
Dalziel, K., Hsu, R.Y., Matthews, B. and Soulie, J.M. (1983)
Biochemistry, in press.
Fatania, H.R., Matthews, B. and Dalziel, K. (1982a) Proc. R. Soc.
London B214 , 369-387.
Fatania, H.R., Matthews, B and Dalziel, K. (1982b) Proc. R. Soc.
London B214, 389-402.
Hsu, R.Y., Mildvan, H.S., Chang, G.G. and Fung, C.H. (1976)J. Biol.
Chem. 251 , 6574-6583.

Hsu, R.Y. and Pry, T.A. (1980) Biochemistry 19, 962-968.
Pry, T.A. and Hsu, R.Y. (1980) Biochemistry 19 , 951-962.
Reynolds, C.H., Kuchel, P.W. and Dalziel, K. (1978) Biochem. J.
171, 733-742.
Reynolds, C.H., Hsu, R.Y. , Matthews, B. , Pry, T.A. and Dalziel, K.
(1978) Arch. Biochem. Biophys. 189 , 309-316.
Vernon, C.M. and Hsu, R.Y. (1983) Arch. Biochem. Biophys. In press.

SECTION III – ENZYME INTERACTIONS AND
DYNAMICS OF ENZYME COMPLEXES

BIOCHEMICAL DYNAMICS IN ORGANIZED STATES : A HOLISTIC APPROACH

G. Rickey Welch

Department of Biological Sciences , University of
New Orleans, New Orleans, Louisiana 70148, USA

"Chemical phenomena in living organisms can never be fully equated with phenomena that take place outside them. This means to say, in other words, that the chemical phenomena of living beings, although they take place according to the general laws of chemistry, always have their own special apparatus and processes". (Bernard, 1878).

INTRODUCTION

The key word in the title of this conference is "dynamics". Indeed, this term comes closest to what we might call the defining characteristic of the "living state". As emphasized so cogently by the eminent biologist, P. Weiss (1973) , we should approach biological systems by reducing them not to their elements of matter, rather to their underlying elementary processes. Ultimately, this reasoning leads us to enzymes, those "chemodynamical machines" (Lumry and Biltonen, 1969) charged teleonomically with the task of catalyzing material transformations at rates (and under appropriate conditions) commensurate with the vitality of the cell.

Enzymology , as a distinct science, is some 100 years old, dating most notably to the work of W. Kühne (Gutfreund, 1976) . Stemming from the first success in isolating and crystallizing enzymes in the early part of this century, as well as from the pioneering kinetic studies by such workers as Michaelis and Menten, enzymology has followed a predictable reductionistic course. This paradigm has carried us far in our attempts to understand the properties of individual enzymes. Yet, enzymology has reached

the point of being consumed in this "grind-and-find" motif
(Wilson, 1980). The central issue is that the microenvironment, in
which the majority of enzymes of intermediary metabolism operate in
vivo, is far different from that in artificial laboratory conditions in vitro . This fundamental distinction was actually discussed
in the early days of enzymology (Bernard, 1978) and was being argued
forcefully as early as the 1930's (Peters , 1930). Unfortunately,
these cries for holism were drowned by the wave of reductionism do-
minating biology throughout the past 30 years or so.

It is our contention, that enzymology has reached a critical
stage in its evolution as a science, at the verge of entering a new
retrograde era of synthesis and integration. The signal for this
"paradigm transition" (Kuhn, 1970) is coming, not from study of the
individual enzyme, but aptly from cell biology itself. In the
present article, we discuss some of the implications of this emer-
ging view on the dynamics of biochemical systems.

MISE EN SCENE

A biphasic picture of cellular infrastructure

In recent decades electron microscopy has revealed a complex
and richly diverse particulate infrastructure in living cells-
especially larger eucaryotic cells. This structure encompasses
the extensive membraneous reticulation (e.g., plasmalemma , endo-
plasmic reticulum, mesosomes [bacterial]), as well as the hyalo-
plasmic space. The latter region, containing the so-called "ground
substance", is laced with a dense array of various cytoskeletal
elements (e.g., microtubules, microfilaments, intermediate fila-
ments) and an interlocking microtrabecular lattice (Clegg, 1984).

Accumulating data indicate that the majority (if not all) of
the enzymes of intermediary metabolism operate in vivo in association with particulate strutures. A perusal of the literature reveals
evidence for enzyme organization in virtually all major metabolic
pathways. The organizational mode may entail formation of protein-
protein complexes and/or individual adsorption to cytological substructures. Strong evidence has come from centrifugation studies
on whole cells (Zalokar, 1960; Kempner and Miller, 1968) and cell
fragments (Coleman, 1970). Experimental and theoretical calculations (Sitte, 1980; Srere, 1981) of protein "concentrations" , in
association with cytomembranes and organelles, indicate high crys-
tal-like densities of protein molecules in (on) particulate structures of the cell. There is a remarkable homology, in the surface·
area-to-volume ratio in all membraneous cytological substructures.
Such considerations led Sitte (1980) to propose, that all cytomem-
braneous elements have evolved in a common fashion to function as
effective "protein collectors" in the operation of cell metabolism.

And, recent work from Porter's group (Schliwa et al, 1981) shows the microtrabecular lattice itself to be "dressed" with a multitude of different proteins, some of which are likely to be metabolic enzymes (Clegg, 1984).

Thus, cell biology presents us with a rather simple, biphasic view of cellular infrastructure: a solid phase, encompassing extensive membrane surfaces and the fibrous lattice-work; and a soluble, aqueous phase (albeit containing a considerable amount of structured water; Clegg, 1984). Empiracle facts and logical netcessity force us to focus on the solid phase as the primary site of intermediary metabolic processes, with the soluble phase functioning largely in such subservient roles as thermal buffering, distribution of common substrates, regulatory substances, and salt ions, etc.

The logic of enzyme organization: cracks in the reductionistic edifice

We find the study of biochemistry at a point in its development, prophesied accurately some 50 years ago by Peters (1930), whereby "in the ultimate structural units of the cell, we have reached the limit for the application of ordinary statistical considerations, and must substitute some more anatomical view, based upon control by surface ... Owing to the microheterogeneous nature of the system, surface effects take precedence over ordinary statistical, mass-action relationships and become in the ultimate limit responsible for the integration of the whole and therefore the direction of activities". Within the confines of the ultimate "metabolic microenvironments" of the living cell-engendered by the aforementioned organizational modes (along with the ambience of structured water, electrical double layers, etc., at the surface boundaries) - we must abandon the concept of scalar chemical reactions, as well as such familiar notions as that of a "uniform concentration". The simplicity, homogeneity, and isotropy of the in vitro conditions are artificial and deceptive. Traditionally, we have defined "enzymes" simply as "proteinaceous catalysts". Now, we must add more biological flavor, with the idea of "locational specificity" part-and-parcel of the defining character (Welch and Keleti, 1981).

In these localized microenvironments, traditional "macroscopic" descriptions of metabolic processes, employing the standard differential equations for reaction/diffusion dynamics, simply break down (Welch, 1977a). In particular, the idea of a "bulk concentration" will not apply, in most instances, in the cellular microenvironments. Here, concentrations of enzymes and their respective substrates are, in many cases, of the same order of magnitude. It is highly plausible, that there are "molecular channels" in orga-

nized multienzyme systems in vivo , wherein each individual enzyme
is subject to a local, "quantized" substrate concentration (Welch,
1977a; Welch and Kell, 1984). A similar"channel" picture has been
proposed for the coupling of proton pumps to H^+-ATPase (Westerhoff
et al, 1983).

Similarly, the "macroscopic" approach is inadequate for depic-
tion of the overall, vectorialized material flow in reaction-diffusion systems exhibiting the kind of inhomogeneity and anisotropy as
those in vivo . We need a physicomathematical construct that will
relate the manner in which the physical components are "hooked-
up". Promising attempts, to date, are seen in the network and
mosaic thermodynamic theories (Westerhoff, 1982). Such mathematical
formalism goes a long way toward describing how things are "hooked-
up", both structurally and functionally, in the intermediary metabolism of the living cell. Yet, something is inherently missing from
the picture; some physical (say, biophysical) "integrative principle"
is lacking. Just how are the elementary processes, in an organized
multienzyme sequence in vivo , coordinated to produce a "sociological unit " ? From holistic inclinations, we intuit that there must
be integrative factors which "zip" an enzyme system into a whole,
which subordinate the individual events (energetically) to the
good of the whole. In short, we need some indication of how metabolism works in the organized state, in addition to our knowledge of
how the components are arranged physically.

Traditionally, we have tended to characterize , thermodynamically, the "organization" of coupled reaction-diffusion processes
in terms of the free-energy flow therein. This course has taken us
far, in our attempts to understand the nature (and chemical "direc-
tionality") of biological processes. However, elucidation of energetic principles, applicable to the microscopic confines of organized states in vivo, demands that the usual thermodynamic analysis be supplemented with molecular details of the functional coupling (McClare, 1974). Without such an information, one is limited
to a "black box" approach - able just to define global parameters.

A BROWNIAN WORLD DISTRIBUTED MAXWELL-BOLTZMANNLY

The motion of a Brownian particle (whether it be due to spatial diffusion or to chemical reactions) can be described by a
Langevin equation, of the form

$$(1) \quad m \frac{dv(t)}{dt} = -\zeta \cdot v(t) + F(r) + R(r,t)$$

where m is the mass, r the spatial coordinate, t the time, v the velocity , ζ the frictional drag on the particle (due to solvent viscosity), F(r) any systematic force on the particle, and R(r,t) the stochastic force arising from individual collisions of the particle with solvent molecules (Kramers, 1940; Kapral, 1981). This famous equation depicts vividly the randomizing influence of a thermalizing environment on particle dynamics. This random field (at a given temperature, T) establishes an equilibrium partitioning of the molecular energy states, according to the well-known Maxwell-Boltzmann relation

(2) $P_i \sim \exp(-E_i/k_B T)$

where P_i is the probability and E_i the energy of the i-th state, and k_B the Boltzmann constant. This relation governs chemical reaction dynamics, as seen from the Arrhenius form for the rate constant, k, as

(3) $k = A. \exp(-E_A/k_B T)$

with E_A the "activation energy" and A a frequency factor.

    Not only does the Maxwell-Boltzmann relation dictate the rate of chemical processes, but also the driving force . Thermodynamically, the rate of a reaction will be some function of the difference in chemical potential (μ_i) between reactants and products.(Strictly, a linear relationship holds only near equilibrium ; Westerhoff, 1982). The chemical potential of molecular species i is defined as

(4) $\mu_i = \mu_i^o + k_B T \ln N_i$

where N_i is the number density of i. Notably, the standard potential μ_i^o relates to the "setting" in which the reaction takes place, being dependent on the Maxwell-Boltzmann energy partition function (McQuarrie, 1976). In the usual thermodynamic formalism, the "driving force" on a chemical reaction (say S $\rightleftharpoons$ P) is specified by the partial free-energy (G) change

(5) $-\left(\dfrac{\partial G}{\partial \xi}\right)_{T,P} = \mu_S - \mu_P$

where ξ is the "extent of reaction". (This partial derivative is sometimes erroneously termed "ΔG" ;Welch and Keleti, 1981)

Life, as an emergent and evolving phenomenon, has had to combat against the random field. Successes in this teleonomic course are evident at all levels of biological complexity. In the remainder of this article, we explore aspects relating to enzymatic rate processes.

CELLULAR DIFFUSION PROCESSES

Regardless of the catalytic capacity of the enzyme, the overall reaction can proceed no faster than diffusion of the substrate to the enzyme. Rendering equ.(1) according to the Smoluchowski theory (McQuarrie, 1976) shows the rate of formation of the enzyme–substrate complex to be proportional to the substrate diffusion coefficient (Eigen and Hammes, 1963). In turn, the diffusion coefficient is inversely proportional to the medium viscosity. Extrapolating directly from in vitro conditions to the living cell, one too often ignores diffusion. Yet, the diffusional impedance of the intracellular milieu is significant, particularly for larger eukaryotic cells. While precise measurements are difficult, it is apparent that the viscosity in cellular microenvironments is rather great (Clegg, 1984; Srere, 1981; Welch, 1977b; Welch et al, 1983a).

A parameter relating to the influence of diffusion on the kinetic behavior of multienzyme sequences is the transient time, which is the time consumed in steady-state transition processes (Gaertner, 1978). This temporal parameter is seen to be proportional to the medium viscosity (Welch , 1977b). Numerical calculation indicates that such transition processes would be unrealistically long, should one regard multienzyme systems as unorganized in vivo (Welch, 1977b).

THE "CYTOSOCIOLOGY" OF ENZYME ACTION IN VIVO

Canvassing the wealth of evidence on the role of enzyme organization in vivo, Keleti (1975) and Welch (1977a) proposed a "cytosociological" view of rate processes in metabolic sequences (Welch and Keleti, 1981). The transition state of enzymic rate processes (see equ. (3)) is the focal point. It seems that enzyme active sites have evolved to be complementary, not to the reactant or product molecular species, rather to the transition-state intermediate between the two. Indeed, this complementarity is at the root of the extraordinary catalytic power of enzymes. This relation befits remarkably the "process view" of living systems espoused above. For, the transition state is, by definition, a molecule (or molecular aggregate) in chemical flux.

Transition-state theory has proven quite useful in enzymology, in dissecting the requisite free energy into separate entropic (e.g., steric) and enthalpic (e.g., energetic) contributions to

a given step in enzyme catalysis. And, discussions of enzyme evolution usually focus on teleonomic modification of the transition-
state barrier for catalytic steps, as well as of the free energy
level ("valley") of the enzyme-substrate complex (Fersht, 1977).
When we embed the isolated enzymatic process in its physiological
role, as an intermediary reaction within the metabolic sequences of
the cell, we find a rich montage of influencing factors which potentially bear upon the unitary rate constants of the isolated process
owing to the "sociological" aspects of enzyme function in vivo.
The physiological advantages of enzyme organization are multifarious,
but they generally fall into two distinct categories. (i) The clus-
tering of the component protein moieties (either among themselves
or with a membraneous matrix) may produce entities that have intrinsic catalytic properties unlike those of the separate proteins,
i.e., the physical association may stabilize and/or enhance the
overall activity of the enzyme sequence.(ii) The assembly into an
organized cluster of enzymes may alter the efficiency of the overall process, even if the intrinsic catalytic activities of the
components are not changed upon association; advantages here result
simply from the proximal juxtaposition of the constituent active
sites within the organized system. In addition to important kinetic
effects, this structuralization may entail unique modes of metabolic
regulation (Keleti, 1978; Welch, 1977a).

As we pry into the functional meshwork of these organized systems, we may learn some valuable lessons on the character of enzyme catalysis in the "living state". Such study is also bound to
heighten the disparity between the conditions imposed on chemical
processes in the cell and those for the corresponding isolated processes in vitro.Accepting the premise that most enzymes of intermediary metabolism operate within some kind of "sociological strcture"
in vivo, we must ultimately "geometrize" enzyme action by construc-
ting appropriate structural rate equations for the corresponding
reaction-diffusion processes (Welch, 1977a). We begin along this
avenue by dissecting the rate constants of enzyme action into a com-
posite form, which depicts the "sociological" influences potentially
brought to bear on the processes represented by these constants.
From this formulation we can relate the influencing factors to the
transition-state theory - which route may yield important clues
(potentially accessible to experimental analysis) as to "energized"
states of matter in organized systems. And, this holistic approach
provides a novel perspective for viewing the energetics of enzyme
evolution relative to in vivo conditions.

For example, consider the(apparent) second-order rate cons-
tant, k_{cat}/K_m , which we abbreviate as k_T (where k_{cat} is the en-
syme catalytic constant and K_m is the Michaelis constant)(Fersht,
1977). We can decompose the (standard) free-energy change for the

corresponding transition state as

(6)
$$\left(\frac{\partial G}{\partial \xi}\right)_{T,P}^{\neq} = \left(\frac{\partial G}{\partial \xi}\right)_{T,P}^{\neq *} + \sum_{i=1}^{n} \left(\frac{\partial G_i}{\partial \xi}\right)_{T,P}^{\neq}$$

where($*$) denotes the <u>intrinsic</u> free-energy change (per molecule) for the transition state in the absence of "external" influences, and the sum on the right-hand side represents the contributions from n extrinsic factors. Then we have

(7)
$$k_T = k_T^* \prod_{i=1}^{n} \alpha_i$$

where

$$k_T^* = A^{\neq}.\exp\left[-\frac{1}{k_B T}\left(\frac{\partial G}{\partial \xi}\right)_{T,P}^{\neq}\right]$$

and

$$\alpha_i = \exp\left[-\frac{1}{k_B T}\left(\frac{\partial G_i}{\partial \xi}\right)_{T,P}^{\neq}\right]$$

Equation (7) has been used (Welch, 1977a) to define the spectrum of extrinsic factors influencing <u>in situ</u> the enzymic rate processes (cf. Keleti, 1975). A marshalling of diverse information, from studies on various physicochemical aspects of enzyme action, suggests that Nature has a repertoire of "cytosociological" factors that can be brought to bear in equ. (7) – owing to physical organization of multienzyme systems (Welch, 1977a; Welch and Keleti, 1981). In addition to the effects on the rate process through equ.(7), some of the organizational features may affect the free-nergy "valley" of the enzyme-substrate complex – which relates to the thermodynamic <u>driving force</u> (viz.$\Delta\mu_i^0$) on the reaction (see equ. (5)).

Underlying equ. (7) is a deeper concept, embodied in the free-energy "complementarity principle" developed by Lumry and coworkers (Lumry and Biltonen, 1969). This "principle" views the protein matrix as a free-energy "buffer", which engages in a fluid and va-

riable exchange with the bound substrate during the course of catalysis (thus "lubricating" the net flow process). The aforementioned
transition-state "barriers" and the free-energy "valleys" (of
stable reactant/product states) define the "ups-and-downs" of cell
metabolism in a Brownian world distributed Maxwell-Boltzmannly.
In a metabolic pathway, the individual enzymes can evolve only so
far in smoothing these "ups-and-downs". Lumry and Biltonen (1969),
in extending their ratiocination to whole metabolic sequences, suggested a "supercomplementarity" applicable to organized clusters of
related enzymes. Whereby, the multienzyme system might be optimized
in evolution as a unit.

     In the insightful work of Lumry, one is confronted inescapably
with the cognition that the enzymological evolutionary battle is
really fought along the free energy vs. reaction - coordinate profile
of the coupled protein-chemical subsystems. And, one realizes that,
within the confines of cellular microenvironments engendered by
enzyme organization, the very chemical potential of "channeled"
chemical substrates (See above) may be inextricably associated with the surrounding ordered matrix (Welch and Keleti, 1981).
(An analogous status rerum may prevail in proton-conduction pathways in energy-transducing membranes (Kell, 1983; Westerhoff et al,
1983)).

DYNAMICS OF THE ENZYME MOLECULE

     Lumry's "complementarity principle" leads us to view the enzyme
as a macromolecular free-energy transducer, with the reactivity of
the protein matrix intimately associated with the enzyme reaction.
Not only does the protein provide the 3-D scaffolding for active-
site processes, but, perhaps more importantly, it serves as the
local solvent for the bound chemical subsystem. Stemming from the
seminal work of such researchers as Linderstrӧm-Lang and Schellman
(1959) and Koshland (1976), it is now apparent that protein motions
are involved in enzyme function. In fact, the old "static" view of
protein structure has become outmoded. Experimental evidence
(Careri et al, 1975) and molecular-dynamics simulation (Karplus and
McCammon, 1983) have revealed a variety of internal motions. Realization of the dynamic nature of the protein molecule has spawned
the development of a number of theoretical models of enzyme action
(Welch et al, 1982; 1983b; Welch, 1984). These models maintain that
particular classes of motions in the enzyme structure generate
high (local) free-energy events at the active site.

     For an enzyme molecule at thermal equilibrium in bulk solution,
the dynamical aspects of catalysis arise from the fluctuational
properties of the protein interacting stochastically with the
solvent (as per equ. 1). Seemingly, this dynamical character must
be sought in slower, collective atomic motions distributed inhomo-

geneously in the protein (Karplus and McCammon, 1983; Doster, 1983; Whittenburg et al, 1984; Welch et al, 1984). Most likely, the lattice structure and dynamics of hydrogen-bond chains are at the root of such collective motions (Lumry and Rosenberg, 1975; Ikegami, 1977).

Viscosity is a key environmental parameter relating to the thermal "(de)energization" of an enzyme-substrate complex (at a given temperature) (Welch et al, 1982). Application of the Fokker-Planck formalism (Gavish, 1978) or fluctuation-dissipation methods (Whittenburg et al, 1984) to equ. (1) shows the enzyme catalytic-turnover constant to be inversely proportional to medium viscosity

$$(8) \qquad\qquad k_{cat} = \left[\tilde{Z}(p) + \frac{1}{\eta_s}\ \tilde{Z}(r)\right] \cdot W(h^{\neq}, T)$$

where η_s is solvent viscosity, $h^{\neq}$ a fluctuation enthalpy for the enzyme reaction, W a probability distribution function (of the Maxwell-Boltzmann form - see equ. 3), and $\tilde{Z}(p)$ and $\tilde{Z}(r)$ are terms dependent on the particular enzyme mechanism (Whittenburg et al, 1984).

Although this thermally-based dynamical view is supported by in vitro evidence, there may be problems when we consider the enzyme molecule in its natural setting in vivo . If we extrapolate the observed viscosity-dependence for enzymes in vitro (following equ. 8) to the kinds of high-viscosity microenvironments extant in vivo, we find a discordant note. Such enzymes would operate in a rather depressed state - assuming thermal "energization" in the modus operandi. For, the internal protein motions would be over-damped (De Brunner and Frauenfelder, 1982; Whittenburg et al, 1984). This consideration raises a serious question about the nature of enzyme action in situ.

"TOWARDS A NEW BIOCHEMISTRY": REVISITED

Szent-Györgyi (1941), in an article of the same title as this section, boldly proposed the first embracing, holistic principle in the context of modern biochemistry. He suggested that electronic semiconduction is an integrative force in the operation of cell metabolism. According to this hypothesis, the machinery of the cell is driven by an "invisible fluid", the particles of which, the electrons, "carry energy, charge , and information, and act as the fuel of life" (Szent-Györgyi, 1968). Over the years, this idea has stimulated much research, particularly into the conductivity properties of proteins. Semiconduction, and other forms of energy

and/or charge transduction (e.g., tunneling, charge-transfer complexes, resonance energy transfer), have been proposed for many
cellular process - most notably, the electron-transfer events
of oxidative- and photo-phosphorylation. Yet, such processes are
restricted to relatively small spatial domains, within (or,in a
few cases, between) individual proteins - usually in the presence of
a membraneous phase. Szent-Györgyi's original idea, viz., the
existence of extensive electron-conduction bands as a general integrative factor in the multienzyme systems of cell metabolism, has
been largely ignored - due primarily to lack of evidence for an
appropriate band-structure from studies on isolated proteins.

Any such conception, as to the existence of long-range energy
continua in the living cell, has not been deemed palatable by the
sciences of biochemistry and enzymology, in their traditional reductionistic posture. Why this stance ? Szent-Györgyi (1979) replied,
that "science has overlooked a very important circumstance", viz.,
the difference between protein function in isolation in vitro and
that in organized states in vivo. This "circumstance" has led to
the emergence, in recent years, of some motive energy-transducing
principles, which may foster a revival in the notion of such long-
range "continua". Permeating subcellular particulates, with which
many enzyme systems are associated (see above), are strong
(local) electric fields and mobile protonic states. Is this juxtaposition fortuitous ? Are the localized enzymatic processes functionally independent of these energy-transducing elements ? Although
hard empirical evidence is presently lacking, there are theoretical
reasons for holding that the "reactivity" of the protein matrix, in
conjunction with the substratum of the aforementioned organizational
modes, may be designed to make full use of these energy sources
(Welch et al, 1984).

Fröhlich (1968, 1983) has developed the notion, that the
activities of localized enzymes couple to electric fields (and
other energy sources) via excitation of metastable states in the
protein. The modality involves longitudinal electric dipolar oscillations of hydrogen-bond units, which are modeled as active phonons
(i.e., quasi-particle lattice vibrations). Such systems can store
external energy in specific modes (and subsequently do work with
it), via a phonon-condensation phenomenon. It has been proposed that
this modality is involved in driving the conformational dynamics of
organized enzyme systems, as well as substrate-translocation processes. Coherent hydrogen-bond phonons (e.g. in α-helical regions),
arising from external energy "pumps", can propagate under some
conditions in the form of dispersion-less wave packets termed
solitons. These soliton modes have been suggested as transduction
devices in the operation of a number of cellular activities (Scott,
1981; Davydov, 1982).

Berry (1981) has offered a novel <u>electrochemical interpreta-</u>
<u>tion of metabolism</u> , which advances the idea that proton current
("proticity"), generated by local redox (or photoredox) processes
and ATP-cleavage reactions, flows through "intracellular circuits"
in structured matrices, specific for various metabolic functions
in the cell. "Structural" proteins in membranes (e,g., "protoneural"
proteins ; Kell, 1983) and microtrabecular lattices, as well as multienzyme systems adsorbed thereto, might sustain this proton trans-
mission. The "proton gradient" is known to interconnect such diverse entities as electron transport, ATP-bond energy, transhydrogena-
tion reactions, mechanical (e.g., flagellar) motion, pH regulation,
active (mass) transport, heat production – and the list is ever-
increasing. Organization of enzyme sequences at cytosol-particulate
interfaces approximates them to sources of such "proton gradients".
The ubiquity of this energy mode may be the most important "integrative" feature of cell metabolism yet discovered.

Recently, Welch and Berry (1983) extended this view, drawing
further attention to a possible physiological connection, between
the central bioenergetic concept of "proticity" <u>and</u> the universal
role(s) of "mobile protons" in enzyme structure, function and
evolution. It was suggested, that organized proton flow in <u>localized</u>
enzyme aggregates can coordinate protochemical, as well as conforma-
tional-dynamical, aspects of enzyme catalysis – in the manner of
"molecular-energy machine" (Welch and Kell, 1984). This approach
embraces the Nagle-Morowitz model of proton-semiconduction in extended hydrogen-bond chains (Nagle and Tristam-Nagle, 1983). Accordingly, such organized enzyme systems are supposed to be "plugged into"
a protonic "continuum" (e.g., at a membrane interface), so that
the action of a given enzyme may be "driven" dynamically by a
steady-state proton flow (in conjunction with local field effects
acting, for example, through α-helix dipoles). This flow would serve,
in somes instances, as a source/sink of protons for protochemical
events at active-centers (Metzler, 1979; Wang, 1968). And, generally, the "energy continuum" of the mobile protonic state, by virtue
of <u>proton-conformational-interaction</u> (Nagle and Tristram-Nagle ,
1983), would serve to"energize" the ES complex (i.e., toward
defined transition-states).

The experimental exploration for long-range energy continua
<u>in vivo</u> is gaining impetus (e.g,, Fröhlich and Kremer, 1983).
Their existence poses major implications, as regards the essence
(and dynamics) of biochemical processes. Natural selection can go
only so far in designing the isolated <u>enzyme molecule</u> to work
against the "random field", especially for enzymes of intermediary
metabolism. Considering the negentropic and anagenetic nature of
biological evolution, we find it difficult to accept the supposition
that metabolic events in the microscopic confines of the cell are
left to pure "chance" . Once the "smoothing" device of the free energy

"complementarity principle" (Lumry and Biltonen, 1969) was pushed to its limits in the evolution of the individual enzyme, perhaps the only recourse was to "plug" many enzymes into such "energy continua" as local electric fields, mobile protonic states, and perhaps others yet to be revealed.

Rate processes for enzyme systems localized in these microenvironments may be quite bizarre, as compared to the familiar Brownian world in vitro. An important message coming from the studies on protein dynamics (see above) is the unity of the protein and the medium (Careri et al, 1975). (Some indication of this "unity" is seen in equ. 8). Under the nonthermal conditions generated by the putative "energy continua", the protein-dynamical modes of enzyme action may be maintained away from equilibrium in specific states, in the sense of a cyclical engine (Welch and Kell, 1984). This may be involved, for example, in the coordination of a whole multienzyme sequence (Fröhlich, 1968; Welch and Berry, 1983).

And, the thermodynamic driving force (see equ. 5) may be very different in these ordered regimes. As alluded above, the structured matrix affects the value of the free-energy change for a chemical reaction, by virtue of its influence on $\Delta\mu_i^0$ (see equ. 4). Moreover, the ambient "energy continuum" (viz. mobile protonic state) may contribute stoichiometrically to certain types of enzyme reactions (e.g., ATPases, kinases, dehydrogenases) involving production or consumption of protons (Nagle and Tristram-Nagle, 1983; Welch and Berry, 1983). Apropos of such systems, Berry (1981) has stressed that our in vitro methods of calculating "ΔG" values may be erroneous. (For example, it is customary to ignore protons involved in our biochemical reactions). This situation puts into disarray, our familiar conceptions of metabolic "efficiency" and metabolic regulation.

Broader implications are realized in the thermodynamic and statistical-mechanical bases of biochemical dynamics. During the last decade or so, the "dissipative structure" construct developed by the Brussels school has placed Biology on a sound thermodynamic foundation (Nicolis and Prigogine, 1977). This paradigm, in elegant fashion, explains how ordered biological systems can emerge out of chaos and develop as an epiphenomenon – driven by dissipation of free energy. Nonetheless, biochemical processes, occurring within the microenvironments of existent subcellular structures, can proceed in a very efficient manner – provided that the intermediary stages of the process are not at the whim of a completely "thermalizing" environment (according to equ.1). This dissipation vs. conservation dichotomy is apparent, not real; for , the two respective characterizations relate to different levels of biological complexity and function.

Biochemical "microprocesses", under the (nonthermal) influence of the long-range energy continua, manifest a <u>quantum-mechanical</u> essence (McClare, 1974; Welch and Kell, 1984). Whereby, we must employ quantum-stastistical energy distributions (as opposed to classical Maxwell-Boltzmann statistics) in equations (2), (3) and (8) (Fröhlich, 1968). This route is looming on the horizon of biochemical dynamics.

In conclusion, we have attempted herein to accentuate the "holistic" aspects of biochemical dynamics. We contend, that a deeper understanding of elementary metabolic processes will require a melding of the following factors : i) the reactivity of the very macromolecular fabric of the enzyme molecule, ii) the material substratum in which most metabolic systems are embedded <u>in vivo</u>, and iii) the essence of motive energy-transducing principles. The first two factors are now founded empirically, as well as theoretically. The third one (as regards "energy continua") remains to be fully established empirically, and it is the underlying "holistic force" in cell metabolism. With the advent of such "bioenergetic" principles as the H^+-electrochemical potential difference (and proton-motive energy coupling), in conjunction with the increasing focus on cytological surfaces as the operational site of most multienzyme systems of cell metabolism, we are forced to infer a much more intimate link between "bioenergetics" and "intermediary metabolism" than heretofore realized.

In the early years of "molecular biology", Szent-Györgyi (1941) mused : "Biochemistry is, at present, in a peculiar state... It looks as if some basic fact about life were still missing, without which any real understanding is impossible". In the intervening years since then the reductionistic emphasis, on the <u>individual</u> processes and the <u>individual</u> enzymes, has served us well. <u>It is a</u> characteristic of <u>science</u>, that analysis must precede synthesis. Hopefully, we are on the right track toward a "real understanding" of biochemistry, using the raw material amassed during the analytical phase. In order to appreciate the "algebra" of holism (viz., 1 + 1 > 2), though, our dialectical course will require us to take "leaps in biochemical logic" (Asensio, 1976) not readily apparent from our <u>in vitro</u> data.

ACKNOWLEDGEMENTS

The author is grateful to the following, for constructive input in the development of ideas presented herein: Drs. M.N. Berry, T. Keleti, D.B. Kell, R. Lumry, B. Somogyi and H.V. Westerhoff.

REFERENCES

Asensio, C. (1976) in Reflections on Biochemistry (Kornberg, A.,
Horecker, B.L., Cornudella, L. and Oro, J. eds.), p.235 , Pergamon,
Oxford.
Bernard, C. (1878) Leçons sur les phénomènes de la vie communs aux
animaux et aux végétaux, translated from the French by H.E. Hoff,
R. Guillemin and L. Guillemin (Charles C. Thomas Publisher,
Springfield, Illinois, 1974).
Berry, M.N. (1981) FEBS Lett. $\underline{134}$, 133-138.
Careri, G., Fasella, P. and Gratton, E. (1975) CRC. Crit. Rev. Bio-
chem. $\underline{3}$, 141-164.
Clegg, J.S. (1984) Amer. J. Physiol., $\underline{246}$, R133-R151.
Coleman, R. (1973) Biochim. Biophys. Acta $\underline{300}$, 1-30.
Davydov, A.S. (1982) Biology and quantum mechanics, Pergamon, Oxford.
De Brunner, P.G. and Frauenfelder, H. (1982) Ann. Rev. Phys. Chem.
$\underline{33}$, 283-299.
Doster, W. (1983) Biophys. Chem. $\underline{17}$, 97-103.
Eigen, M. and Hammes, G. (1963) Adv. Enzymol. $\underline{25}$, 1-38.
Fersht, A. (1977) Enzyme structure and mechanism, Freeman, San
Francisco.
Fröhlich,H. (1968) Int. J. Quant. Chem. $\underline{2}$, 641-649.
Fröhlich,H. and Kremer, F. (eds.) (1983) Coherent excitation in
biological systems, Springer, Heidelberg.
Gaertner, F.H. (1978) Trends Biochem. Sci. $\underline{3}$, 63-65.
Gavish, B. (1979) Biophys. Struct. Mech. $\underline{4}$, 37-52.
Gutfreund, H. (1976) FEBS Lett. $\underline{62}$ (Supplement) , E1-E2.
Ikegami, A. (1977) Biophys. Chem. $\underline{6}$, 117-130.
Kapral, R. (1981) Adv. Chem. Phys. $\underline{48}$, 71-181.
Karplus, M. and McCammon, J.A. (1983) Ann. Rev. Biochem. $\underline{53}$, 263-
300.
Keleti, T. (1975) in Proc. Ninth. FEBS Meeting, Vol. 32 (Symposium
on mechanism of action and regulation of enzymes) (Keleti, T. ed.),
p.3, North-Holland, Amsterdam.
Keleti, T. (1978) in New trends in the description of the general
mechanism and regulation of enzymes (Damjanovich, S., Elodi, P.
and Somogyi, B. eds.), p.107, Akademiai Kiado, Budapest.
Kell, D.B. (1983) in Coherent excitations in biological systems
(Fröhlich, H. and Kremer, F., eds.), p.205, Springer, Heidelberg.
Kempner, E.S. and Miller, J.H. (1968) Exp. Cell. Res. $\underline{51}$, 150-156.
Koshland, D.E.Jr. (1976) FEBS Lett. $\underline{62}$ (Supplement) ,E47-E52.
Kramers, H.A. (1940) Physica $\underline{7}$, 284-304.
Kuhn, T. (1970) The structure of scientific revolutions (2nd ed.)
University of Chicago Press, Chicago.
Linderström-Lang, K.U. and Schellman, J.A. (1959) in The enzymes,
Vol 1 (2 nd ed.)(Boyer, P.D., Lardy, H.and Myrback, K. eds.),
p.443, Academic Press, New-York.

Lumry, R. and Biltonen, R. (1969) in Structure and stability of biological macromolecules (Timasheff, S.N. and Fasman, G.D., eds.) p.65, Dekker, New-York.

Lumry, R. and Rosenberg, A. (1975) Colloques Internationaux du C.N.R.S. (n°246, "L'eau et les systèmes biologiques"), pp.53-62.

McClare, C.W.F. (1974) Ann. N-Y. Acad. Sci 227, 74-97.

McQuarrie, D.A. (1976) Statistical mechanics, Harper and Row, New-York.

Metzler, D.E. (1979) Adv. Enzymol. 50, 1-40.

Nagle, J.F. and Tritram-Nagle, S. (1983) J. Memb. Biol. 74, 1-14.

Nicolis, G. and Prigogine, I. (1977) Self-organization in non-equilibrium systems, Wiley, New-York.

Peters, R.A. (1930) Trans. Farad. Soc. 26, 797-807.

Schliwa, M., van Blerkom, J. and Porter, K.R. (1981) Proc. Nat. Acad. Sci. Usa 78, 4329-4333.

Scott,A.C. (1981) in Nonlinear phenomena in physics and biology (Enns, R.N., Jones, B.L., Miura, R.M. and Rangnekar, SS. eds.), p.7, Plenum, New-York.

Sitte, P. (1980) in Cell compartmentation and metabolic channelling (Nover, L., Lynen, F. and Mothes, K., eds.) p.17, Elsevier/North-Holland, New-York.

Srere, P. (1981) Trends Biochem. Sci. 6, 4-7.

Szent-Györgyi, A. (1941) Nature (London) 148, 157-159.

Szent-Györgyi, A. (1968) Bioelectronics, Academic Press , New-York.

Szent-Györgyi, A. (1970) in Submolecular biology and cancer (Ciba foundation Symposium n°67), Excerpta Medica, New-York.

Wang, J.H. (1968) Science 161, 328-334.

Weiss, P. (1973) The science of life, Futura, Mount Kisko, New-York.

Welch, G.R. (1977a) Prog. Biophys. Mol. Biol. 32, 103-191.

Welch, G.R. (1977b) J. Theor. Biol. 68, 267-291.

Welch, G.R.(ed) (1984) The fluctuating enzyme, Wiley, New York, to appear.

Welch, G.R. and Berry, M.N. (1983) in Coherent excitations in biological systems (Fröhlich, H. and Kremer, F., eds.) p.95, Springer, Heidelberg.

Welch, G.R. and Keleti, T. (1981) J. Theor. Biol. 93, 701-735.

Welch, G.R. and Kell, D.B. (1984) "Not just catalysts-Molecular machines in bioenergetics" to appear in The fluctuating enzyme (Welch, G.R., ed.) , Wiley, New-York.

Welch, G.R., Somogyi, B. and Damjanovich, S. (1982) Prog. Biophys. Mol. Biol. 39, 109-146.

Welch, G.R., Somogyi, B., Matko, J. and Papp, S. (1983) J. Theor. Biol. 100, 211-238.

Welch, G.R., Somogyi, B. and Damajanovich, S. (1984) Biochim. Biophys. Acta (Reviews on Bioenergetics), in press.

Westerhoff, H.V. (1982) Trends Biochem. Sci. 7, 275-279.

Westerhoff, H.V., Melandri, B.A., Venturoli, G., Azzone, G.F. and Kell, D.B. (1984) FEBS Lett. 165, 1-5.

Whittenburg, S.L. , Wan, W. and Welch, G.R. (1984) "on the protein
dynamics of enzyme action : role of viscosity " , submitted.
Wilson, J.E. (1980) Curr. Top. Cell. Regul. 16, 1-54.
Zalokar, M. (1960) Exp. Cell. Res. 19, 114-132.

CHANNELLING IN ENZYME COMPLEXES

Tamàs Keleti

Insitute of Enzymology, Biological Research Center
Hungarian Academy of Sciences, Budapest Hungary

INTRODUCTION

The main goal to be reached by an enzymologist is to understand the unity of structure and function of enzymes in vivo , i.e. the knowledge of the organization, localization of enzymes and substrates, the functioning of enzymes and the regulation of metabolic pathways in the cell. In the cytoplasm a multitude of polyelectrolyte macromolecules and of small molecules coexist. If the overall concentration of substrates, etc. would reach the level which is optimal for the enzymes serious solubility problems would arise in the cell (Atkinson, 1969). Consequently, the conservation of low overall but high local concentrations of metabolites is crucial to the vitality of the cell. The ability of enzymes to work efficiently at low overall concentrations of intermediates is of great importance energetically, i.e. for the economy of the cell. The formation of channels, i.e. microcompartments able to compartmentalize intermediates thus assuring the local high concentration of substrates for the optimal functioning of enzymes, may occur through dynamic compartmentation of the enzymes forming "complementary cages" by the juxtaposed active centers (Friedrich, 1974). There are already a number of examples on multienzyme complexes formed in the cell and demonstrated in vitro and somewhere also the channelling effect has been proved (Welch, 1984).

One of the best known multienzyme complexes is the tryptophan synthase of E. coli consisting of 2α and 2β subunits. The α and β subunits mutually activate each other by the formation of the multienzyme complex (Hatanaka et al, 1962; Wilson and Crawford, 1965) and mutually influence each other's substrate-binding pro-

perties (Kirschner et al, 1975, 1975a). During the catalytic
reaction the intermediate indole is not liberated into the bulk
medium but is channelled between the α and β subunits (Yanofsky
and Crawford, 1972), which was shown for the enzyme isolated from
N. crassa (Matchett, 1974) and yeast (Manney, 1970).

Another well-known multienzyme complex is the pyruvate dehydrogenase of E. coli consisting of pyruvate decarboxylase, lipoate
acetyltransferase and lipoate dehydrogenase , which forms a shunt
between glycolysis and the Krebs cycle. The reactions are coupled
and no intermediate is released in the bulk medium (Bata et al,
1977).A similar mechanism was found with the mammalian enzyme
complex (Cate et al, 1980).

The third widely investigated multienzyme complex is the
α-ketoglutarate dehydrogenase complex, catalyzing a part of the
Krebs cycle. The enzymes from E. coli and from mammalian sources
show a similar channelling mechanism as the pyruvate dehydrogenase
(Collins and Reed, 1977 ; Severin et al, 1978). However , in these
two cases one cannot speak of real channelling, since the intermediate is temporarily covalently linked to the enzyme.

The conjugate enzyme complex of the biosynthesis of phenylala-
nine, tyrosine and tryptophan, the "arom" multienzyme complex
isolated from N. crassa, contains five different enzyme activities
in a single polypeptide chain (Gaertner and Cole, 1977). Due to the
channelling of the intermediates the transient time (i.e. the
time needed to reach the steady state of the coupled system) of the
five reaction steps from the first substrate up to the end-product
is about tenfold lower than that of.a hypothetical mixture of the
five independent (physically unassociated) enzymes catalyzing the
same reactions with the same kinetic parameters (Welch and Gaertner,
1975; Gaertner et al, 1970).

A two-enzyme conjugate of orotate phosphoribosyl transferase and
orotidine-5'-P decarboxylase from Ehrlich ascites cells channels
the intermediate orotidine-5'-P (Traut and Jones, 1977; Traut,
1980).

Not only tight enzyme complexes and enzyme conjugates may
show channelling effect, but also weak enzyme complexes which can
be formed between some cytosolic, so called "soluble" , enzymes.
In the glycolytic pathway aldolase and D-glyceraldehyde-3-phospha-
te dehydrogenase isolated from mammalian muscle were found to form
a weak complex (Keleti et al, 1977; Ovàdi et al, 1978; Grazi and
Trombetta, 1980) and kinetic and thermodynamic evidences for the
channelling of glyceraldehyde-3-phosphate intermediate have been
presented (Ovàdi and Keleti, 1978; Patthy and Vas, 1978; Keleti,
1978).

Channelling was manifest also in the "glycosome" composed of the glycolytic enzymes and found in the lysate of E. coli (Mowbray and Moses, 1976; Gorringe and Moses, 1980).

Channelling of metabolites supposedly occurs in the complex of pig heart citrate synthase and malate dehydrogenase (Halper and Srere, 1977; Srere, 1976).

The interaction of mitochondrial or cytoplasmic malate dehydrogenase and aspartate aminotransferase (Backman and Johansson, 1976) results in the virtual disappearance of transient time, which is an indication of channelling of the intermediate, oxaloacetate (Bryce et al, 1976).

The complex formation of glutamate dehydrogenase and aspartate aminotransferase (Fahien and Smith , 1974; Salerno et al, 1975) results in the shortening of the transient time, also suggestive of channelling of the intermediate (Salerno et al, 1982).

As far as the methodology to test the existence of channelling is concerned a great variety of qualitative and quantitative methods is at our disposal.

ANALYSIS FOR THE TRANSIENT TIME IN COUPLED REACTION

In a coupled reaction of two enzymes, E_1 and E_2, designating the substrate of E_1 by S, its product , the intermediate, which is the substrate of E_2, by I and the product of E_2, the end-product of the coupled reaction by P, we have the following sequence of reactions :

$$S \xrightarrow{\quad E_1 \quad} I \xrightarrow{\quad E_2 \quad} P$$

For a simple way of analysis of the reaction it is necessary to assure the reaction of E_1 to be zero order with respect of S, i.e. $[S]$ should be high. The reaction of E_2 should be first order with respect of I, i.e. $[I]_{ss} \ll K_{M,E_2}$, where subscript ss denotes the steady state concentration and K_{M,E_2} means the Michaelis constant for enzyme E_2. We denote with k_1 and k_2 the first order rate constants of the rate limiting steps of the reactions catalyzed by E_1 and E_2 , respectively assuming that both enzymes follow Michaelis-Menten mechanism (cf. Keleti, 1981).

The production of P reaches the steady state only after a lag period, the transient time : τ (Barwell and Hess, 1970; Hess and Wurster, 1970). The steady state region of the plot $[P]$ vs. t is approximately linear and its extrapolation intersects the abscissa at τ, and the ordinate at $-[I]_{ss}$. The slope of this straight line equals the steady state rate of the reaction catalyzed by E_1, i.e. v_1. Since the reaction of E_1 should be of zero order with respect to S :

$$(1) \qquad v_1 = k_1 [E_1]$$

In the steady state

$$(2) \qquad [P] = v_1(t - \tau)$$

From the law of mass conservation (if at t=0, $[P] = [I] = 0$) it follows that

$$(3) \qquad [P] = v_1 t - [I]_{ss} - v_2/k_2$$

Consequently

$$(4) \qquad v_1 \tau = [I]_{ss} + v_2/k_2$$

In the steady state $v_1 = v_2$ and since

$$(5) \qquad v_2 = V_{max,E_2} [I]_{ss}/K_{M,E_2}$$

we obtain (Bartha and Keleti, 1979) that

$$(6) \qquad \tau = (K_{M,E_2} + [E_2])/k_2[E_2]$$

If $[E_2] \ll K_{M,E_2}$

$$(7) \qquad \tau = K_{M,E_2}/k_2 [E_2] \qquad \text{(cf. Hess and Wurster, 1970)}$$

$1/\tau$ is a linear function of $[E_2]$ and it is independent of $[E_1]$. If $1/\tau$ is not a linear function of E_2 (where $[E_2] \ll K_{M,E_2}$) the complex formation of E_1 and E_2 is probable.

If K_{M,E_2} or V_{max,E_2} as detected from the transient time of the coupled reaction differs from that measured with E_2 alone, channelling is to be assumed (Ovádi and Keleti, 1978; Bartha and Keleti, 1979). Namely , first the kinetic parameters of E_2 (i.e. k_2' and K_{M,E_2}') should be determined in the absence of E_1. Then the same kinetic parameters (denoted k_2'' and K_{M,E_2}'') should be measured by adding the substrate of E_2 to E_2 in the presence of E_1 (non-functioning enzyme, which lowers the functioning free E_2 concentration if a chanelling complex is formed). If $k_2' = k_2''$ and $K_{M,E_2}' = K_{M,E_2}''$, the enzyme E_1 is unable to interact with E_2 or this interaction has no kinetic consequence. If the two equalities do not hold, the reverse statement is valid. The third step of the analysis is the determination of the kinetic parameters of E_2 in the coupled system (k_2 and $K_{M,E2}$) . The analysis can be performed where $[E_2]_0 \ll [I]_{ss}$, but in this case $[I]_{ss} \ll K_{M,E2}$ is not a necessary condition if $[S]_0$ is high enough to ensure that v_1 = constant. If a channelling complex exists $k_2'' \neq k_2$ or $K_{M,E2}'' \neq K_{M,E2}$.

ANALYSIS OF THE COUPLED REACTION IN THE PRESENCE OF ADDED INTERMEDIATE.

Further information on the existence of a channelling complex can be obtained by determining the substrate concentration dependence of the steady state rate of the coupled system with the substrate of E_2 added (S_2 which by definition equals I). In this case we analyze the following reaction (Bartha and Keleti, 1979):

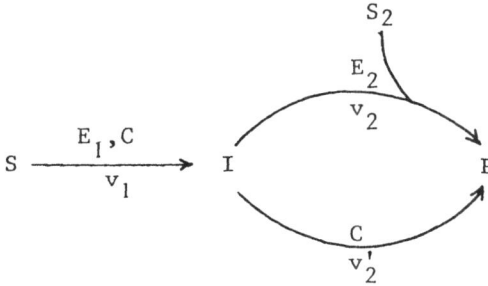

Scheme A

where C is the complex of E_1S (E_1 saturated with S) and E_2, v_1 is the steady state rate of E_1 reaction , and is assumed to equal that of C (i.e. E_2 does not influence the activity of E_1), E_2 catalyzes the conversion of I with rate v_2, while C with v_2', i.e. E_1 influences the activity of E_2. We denote $[S''] = [S_2] + [I]$

(8) $[S''] = [S_2]_0 + v_1 \tau$

One must make sure that S_2 is not a modifier of E_1 and $[S_2]$ is not so high to change the activity of E_1 either by the reversibility of the reaction or by product modulation.

For the qualitative determination of interaction and/or channelling two simplified versions of Scheme A are of interest. Assuming E_2 = C = E_2' (i.e. kinetically E_2 and C are indistinguishable, since E_1 does not influence the activity of E_2) :

$$E_2' + S_2 \underset{k_{-1}}{\overset{k_1}{\rightleftharpoons}} E_2'S_2 \xrightarrow{k_2'} E_2' + P$$

$$E_2' + I \underset{k_{-1}'}{\overset{k_1'}{\rightleftharpoons}} E_2'I \xrightarrow{k_2''} E_2' + P$$

Scheme B

In Scheme B $K_M' = (k_{-1} + k_2')/k_1$ and $K_M'' = (k_{-1}' + k_2'')/k_1'$. If $k_2' = k_2''$ channelling occurs, since kinetically S_2 is distinguished from I. S_2 and I "compete" for E_2' but the competition depends on their physical segregation rather than on the ratio of their concentrations and affinities. The reciprocal reaction rate of "two" substrates catalyzed by the "same" enzyme and leading to the same product is generally not linear as a function of reciprocal total substrate (S") concentration. The channelling effect implies that $K_M' \neq K_M''$, but the reverse is not true. The above inequality may indicate channelling since the substrate saturation curve, from which K_{M,E_2}'' is calculated, is constructed with the average concentration (taking into account the total volume of the bulk medium), but the enzyme encounters a much higher substrate concentration prevailing in the channel.

The second simplified version of Scheme A is presented in Scheme C.

$$E_2 + S'' \underset{k_{-1}}{\overset{k_1}{\rightleftharpoons}} E_2S'' \xrightarrow{k_2'} E_2 + P$$

$$C + S'' \underset{k_{-1}'}{\overset{k_1'}{\rightleftharpoons}} CS'' \xrightarrow{k_2''} C + P$$

Scheme C.

Scheme C involves the assumption that the free enzyme, E_2, and the complex, C, are unable to differentiate between I and S_2. This means that interaction without channelling is implied since kinetically $E_2 \neq C$, whereas $S_2 = I$ not only chemically but also kinetically due to the lack of physical segregation. The reciprocal rate of reactions with the same substrate catalyzed by "two" enzymes and yielding the same product is generally not linear as a function of reciprocal total substrate (S") concentration.

If $k_2' = 0$ and $K_M' = \infty$ (complete channelling, assuming complete complex formation) for Scheme B:

(9) $$v_\theta = k_2'' \, [I] \, [E_2]_0 \, / (K_M'' + [I] \,)$$

and for Scheme C

(10) $$v_0 = k_2'' \, [S''] \, [E_2]_0 / (K_M'' + [S''] \,)$$

In the case B and C, v_θ is independent of $[S_2]$ (depends only on $[S_2] + [I] = [S'']$ in case C), and $1/v_0$ vs. $1/[S'']$ is not linear if Scheme B holds, but is linear if Scheme C were valid. However, the latter cannot hold true since Scheme C implies no channelling.

The general relationships, which in some cases allow one to differentiate between interaction, channelling or their absence, are summarized in Table 1.

TRAPPING METHODS

The common feature of trapping methods is the lowering of the concentration of the intermediate for the second enzyme in the coupled reaction by physical segregation, chemical reaction or enzymatic degradation unless this is prevented by channelling.

Trapping by physical segregation

By using a solvent which does not affect the activity of either the first or the second enzyme, in which the intermediate is soluble but the substrate and the end-product are not, one can prove or disprove the channelling effect. Yanofsky and Rachmeler (1958) used toluene as trapping agent to prove the channelling of indole intermediate of tryptophan synthase multienzyme complex from N. crassa .

Trapping by chemical reaction

By using any compound which does not affect the activity of the enzyme in the coupled reaction and does not react with the substrate and the end-product but reacts with the intermediate, one can prove or disprove the existence of a channel. The simplest reactant is water .E.g. D-glyceraldehyde-3-phosphate, in aqueous solution exists in two forms : the aldehyde and the diol (hydrated) form. The enzymes produce and use only the aldehyde form of the triosephosphate but in water the equilibrium is shifted towards the diol form. Ovàdi and Keleti (1978), by demonstrating kinetically the absence of the diol form of glyceraldehyde-3-phosphate in the coupled reaction of aldolase and D-glyceraldehyde-3-phosphate dehydrogenase, proved that the intermediate did not mix with the bulk

Table 1. Various forms of equation $v_0 = f([S''])$ and the shape of $1/v_0$ vs. $1/[S'']$

Model	$k_2' \neq k_2''$, $K_M' \neq K_M''$		$k_2' \neq k_2''$, $K_M' = K_M'' = K$	
	equation	shape	equation	shape
Scheme B	$v_0 = k_2''(b[S_2] + a[I])[E_2']_0/(K_M' + a[I] + [S_2])$	non linear[+]	$v_0 = k_2''(b[S_2] + [I])[E_2']_0/(K + [I] + [S_2])$	non linear[++]
Scheme C	$v_0 = k_2''[E_2]_0 \left[[S'']/([S''] + K_M'') + b[S'']/([S''] + aK_M'')\right]$	non linear[++]	$v_0 = k_2''[E_2]_0 \left[[S''](1+b)/([S''] + K)\right]$	linear[++]

Model	$k_2' = k_2'' = k$, $K_M' = K_M'' = K$		$k_2' = k_2'' = k$, $K_M' \neq K_M''$	
	equation	shape	equation	shape
Scheme B	$v_0 = ([S_2] + [I])k[E_2']_0/(K + [S_2] + [I])$	linear[□]	$v_0 = ([S_2] + a[I])k[E_2']_0/(K_M' + a[I] + [S_2])$	non linear[§]
Scheme C	$v_0 = 2k[E_2]_0 \left[[S'']/([S''] + K)\right]$	linear[□]	$v_0 = k[E_2]_0 \left[[S'']/([S''] + K_M') + [S'']/([S''] + aK_M'')\right]$	non linear[++]

$K_M'/K_M'' = a$, $k_2'/k_2'' = b$

[+] Interaction with or without channelling
[§] Channelling with or witout other interaction
[□] No Interaction, no channelling
[++] Interaction without channelling.

medium (water) thus demonstrating the existence of a channel between the active centers of the two enzymes. Similarly, Keleti (1978) using a batch microcalorimeter demonstrated the absence of the heat of reaction of imine formation between glyceraldehyde-3-phosphate and tris buffer, the in vitro medium.

Trapping by enzyme reaction

A specific enzyme may transform the intermediate to a compound which no longer is a substrate of the second enzyme if the intermediate is not protected by channelling. This "enzyme probe" method elaborated by Friedrich (Friedrich et al , 1977; Solti and Friedrich, 1979) gave evidence of metabolite compartmentalization (channelling) of a part of the glycolytic pathway in the erythrocyte, namely that of triose phosphates and NAD.

DETERMINATION OF CHANNELLING BY ISOTOPE DILUTION

Addition of isotopic intermediate to the coupled reaction

By adding isotopic intermediate to the coupled reaction system working with non-isotope substrate, one can determine the specific activity of the isolated end-product. In the case of complete channelling the specific activity of the end-product will be zero. In the total absence of channelling the specific activity of end-product will equal that of the added intermediate diluted with the intermediate produced in the coupled reaction. In the case of partial channelling the isotope labelling is intermediate and inversely proportional to the degree of channelling (Friedrich, 1984).

Adding isotopic substrate and non-labelled intermediate to the coupled reaction

The other possibility is to start the coupled reaction with isotopic substrate (S*) and to measure the specific radioactivity of the end-product (P) . Since

$$S^* \xrightarrow{\quad v_1 \quad} I^* \xrightarrow{\quad k_2 \quad} P^*$$

we have

(11)
$$[P] = v_1 t + (v_1/k_2)(e^{-k_2 t} - 1)$$

By diluting the system with non-labelled I, we can calculate the specific radioactivity (r) of P (Bryce et al, 1976). If no channelling occurs

$$(12) \quad r_1 = (v_1 t + (v_1/k_2)(e^{-k_2 t} - 1))/(v_1 t + (v_1/k_2)(e^{-k_2 t} - 1) + [S^*](1 - e^{-k_2 t}))$$

In the case of partial channelling

$$(13) \qquad\qquad r_2 = v_1 t/(v_1 t + [S^*](1 - e^{-k_2 t}))$$

and if complete channelling is detected $r_3 = 1$.

It is ever more understood that Nature evolved the enzyme complexes and conjugates to avoid the loss, dilution and side reactions of metabolites thus sparing energy in the better adapting species (Keleti, 1975; Welch, 1977; Welch and Keleti, 1981). Consequently the channelling and compartmentalization of intermediates is probably a general phenomenon which merits much more attention in the future. We believe that in the years to come the importance of the in vitro demonstration of enzyme complexes, of metabolite compartmentalization and of the channelling effect will be appreciated.

REFERENCES

Atkinson, D.E. (1969) Curr. Top. Cell. Reg. 1, 29-43.
Backman, L., Johansson, G. (1976) FEBS Lett. 65, 39-43.
Bartha, F., Keleti, T. (1979) Oxid. Commun. 1, 75-84.
Barwell, C.J., Hess, B. (1970) Hoppe Seyler's Z. Physiol. Chem. 351, 1531-1536.
Bata, D.L., Danson, M.J., Hale, G., Hooper, E.A., Perham, R.N. (1977) Nature, 268 , 313-316.
Bryce, C.F.A. , Williams, D.C., John, R.A., Fasella, P. (1976) Biochem.J. 153 , 571-577.
Cate, R.L., Roche, T.E., Davis, L.C. (1980) J. Biol. Chem. 255, 7556-7562.
Collins, J.H., Reed, L.J. (1977) Proc. Natl. Acad. Sci. USA 74, 4223-4227.
Fahien, L.A., Smith, S.E. (1974) J. Biol. Chem. 249 , 2696-2703.
Friedrich, P. (1974) Acta Biochim. Biophys. Acad. Sci. Hung. 9, 159-173.
Friedrich, P. (1984) Enzymes: Quaternary structure and beyond. Akadémiai Kiadó, Budapest.
Friedrich, P., Apró-Kovács, A.V., Solti, M. (1977) FEBS Lett. 84, 183-186.
Gaertner, F.H. , Cole, K.W. (1977) Biochim. Biophys. Res. Commun. 75, 259-264.
Gaertner, F.H., Ericson, M.S., DeMoss, J.A. (1970) J. Biol. Chem. 245, 595-600.

Gorringe, D.M., Moses, V. (1980) Int. J. Biol. Macromol. 2, 161-173.

Grazi, E., Trombetta, G. (1980) Eur. J. Biochem. 107, 369-373.

Halper, L.A., Srere, P.A. (1977) Arch. Biochem. Biophys. 184, 529-534.

Hatanaka, M., White, E.A., Horibata, K., Crawford, I.P. (1962) Arch. Biochem. Biophys. 97 , 596-606.

Hess, B., Wurster, B (1970) FEBS Lett. 9, 73-77.

Keleti, T. (1975) in Mechanism of Action and Regulation of Enzymes (ed. T. Keleti) 9th FEBS Meeting . Vol. 32. Akadémiai Kiadó, Budapest. pp. 3-27.

Keleti, T.(1978) in New Trends in the Description of the General Mechanism and Regulation of Enzymes (Eds. S. Damjanovich, P. Elödi, B. Somogyi). Symp. Biol. Hung. 21. Akadémiai Kiadó, Budapest. pp. 107-130.

Keleti, T. (1981) in Kinetic Data Analysis (Ed. L. Endrényi) Plenum Press, New-York. pp 353-374.

Keleti, T., Batke, J., Ovàdi, J., Jancsik, V., Bartha, F . (1977) Adv. Enzyme Regul. 15 , 233-265.

Kirschner, K., Weischet, W., Wiskocil, R.L. (1975) in Protein-Ligand Interaction (Eds. H. Sund, G. Blauer) Walter de Gruyter, Berlin. pp. 27-44.

Kirschner, K., Wiskocil, R.L., Foehn, M., Rezeau, L. (1975a) Eur. J. Biochem. 60, 513-523.

Manney, T.R. (1970) J. Bacteriol. 102 , 483-488.

Matchett, W.H. (1974) J. Biol. Chem. 249, 4041-4049.

Mowbray, J., Moses, V. (1976) Eur. J. Biochem. 66 , 25-36.

Ovàdi, J., Keleti, T. (1978) Eur. J. Biochem. 85, 157-161.

Ovàdi, J., Salerno, C., Keleti, T. , Fasella, P. (1978) Eur. J. Biochem. 90, 499-503.

Patthy, L. , Vas, M. (1978) Nature 276, 94-95.

Salerno, C., Ovàdi, J., Churchich, J., Fasella, P. (1975) in Mechanism of Action and Regulation of Enzymes (Ed. T. Keleti) Akadémiai Kiádo, Budapest. pp. 147-160.

Salerno, C., Ovàdi, J., Keleti, T., Fasella, P. (1982) Eur. J. Biochem. 121 , 511-517.

Severin, S.E., Gomazkova, V.S., Krasovskaya, O.E., Stafeeva, O.A. (1978) Biokhimiya 43, 2241-2248.

Solti, M., Friedrich, P. (1979) Eur. J. Biochem. 95 , 551-559.

Srere, P.A. (1976) in Gluconeogenesis : Its Regulation in Mammalian Species (Eds. R.W. Hanson, W.A. Mehlman) John Wiley and Son, New-York. pp. 153-161.

Traut, T.W. (1980) Arch. Biochem. Biophys. 200, 590-594.

Traut, T.W. , Jones, M.E. (1977) J. Biol. Chem. 252, 8374-8381.

Welch, G.R. (1977) Progr. Biophys. Mol. Biol. 32 , 103-191.

Welch, G.R., Gaertner, F.H. (1975) Proc. Natl. Acad. Sci. USA 72, 4218-4222.

Welch, G.R. and Keleti, T. (1981) J. Theor. Biol. 93, 701-735.
Welch, G.R. (1984) This volume.
Wilson, D.A. and Crawford, I.P. (1965) J. Biol. Chem. 240, 4801-4808.
Yanofsky, C. and Crawford, I.P. (1972) in The Enzymes (Ed. P. Boyer)
Vol. 7, Acad. Press. New-York, London, 3rd ed. pp. 1-31.
Yanofsky, C. and Rachmeler, M. (1958) Biochim. Biophys. Acta 28, 640-645.

CONFORMATION CHANGES IN THE ASSEMBLY OF THE $\alpha_2\beta_2$

COMPLEX OF TRYPTOPHAN SYNTHASE

A.N. Lane, C.H. Paul and K. Kirschner

Department of Biophysical Chemistry, Biozentrum
University of Basel, CH 4056 Basel, Switzerland

INTRODUCTION

Many studies have addressed the question why most soluble enzymes are oligomeric (Welch, 1977, Friedman and Beychock, 1979; Jaenicke, 1982). By investigating the kinetics of refolding and reassociation of completely denatured proteins, and also testing the properties of folded but immobilized monomers it has been shown that assembly often modifies the intrinsic catalytic properties of monomers. In the extreme, monomers have very low enzymic activity by themselves and therefore must be activated in the course of the assembly process. The interpretation usually given for this phenomenon is a change in monomer conformation induced by subunit-subunit interactions, but direct evidence is scarce. In most cases experiments are difficult because identical monomers have high affinity for each other, which precludes physico-chemical studies of folded monomers at high concentrations in solution.

In contrast, oligomeric multi-enzyme complexes can be separated into monomers, because they are chemically different. This is illustrated by haemoglobin A (Tainsky and Edelstein, 1973) and ATCase (Bothwell and Schachman, 1980). Thus it is feasible to measure the rates of assembly from folded, isolated monomers by rapid reaction techniques, and to elucidate the assembly mechanism at the level of elementary steps. In this manner one should be able to distinguish between second-order reactions in which initial complexes are formed on one hand, and first order conformation changes on the other.

Unfortunately, the rate of assembly of haemoglobin A is determined by the rate of dissociation of β_4 to β monomers (Mc Donald , 1981), resulting in only a single measurable rate process. The complex

assembly mechanism of ATCase is described to date only by apparent
second-order rate constants, because rapid mixing experiments have
not yet been performed (Bothwell and Schachman, 1980). Tryptophan
synthase, an $\alpha_2\beta_2$ complex from <u>Escherichia coli</u> (Yanofsky and
Crawford, 1972; Miles, 1979) , turns out to be a useful assembly
system for answering experimentally (and as directly as possible)
the following questions : (i) Can one detect first-order processes,
which may be related to subunit conformation, and (ii) , what role,
if any, do isomerization processes play in changing the intrinsic
properties of the subunits during assembly ? Here we describe rapid
mixing experiments, which reveal that the assembly of tryptophan
synthase involves conformation changes of both subunits.

      Tryptophan synthase is readily dissociated into two α subunits
and one tightly associated β_2 subunit. The turnover numbers of
both the A reaction catalyzed by the α subunit and the B reaction
catalyzed by the β protomer are enhanced approximately 100-fold
after forming the $\alpha_2\beta_2$ complex (Yanofsky and Crawford, 1972; Miles,
1979). These effects have been used to measure the kinetics and
thermodynamics of assembly (Creighton and Yanofsky, 1966; Miles,
1970; Mosteller and Goldstein, 1977).However, these studies necessarily require the presence of substrates, which in turn affect
the assembly equilibrium. Moreover, this approach is not suited
for detecting the elementary steps of the assembly process. The
coenzyme pyridoxal phosphate, which is tightly bound to the active
site of the β protomer, responds to the binding of the α subunit
(Kirschner et al, 1975), and is therefore a convenient natural
reporter group for the desired kinetic studies.

RESULTS

      How tightly is the α subunit bound to the β_2 subunit ?
Spectrofluorimetric titration at the lowest concentrations feasible
renders only a sharp break in the fluorescence increase at the equivalence point (Figure 1). The upper limit of $K_d = 10^{-9}$M, and the
binding of each α subunit increases the fluorescence to the same
degree.

      Are the binding sites for the α subunit equivalent ? The
answer is no, at least when phosphate buffer pH 7.6 is used.
Adachi et al (1974) had previously shown that the $\alpha\beta_2$ subcomplex is
a stable   species, which does not disproportionate to $\alpha_2\beta_2 + \beta_2$.
This observation means that the first α subunit is bound more
tightly than the second. In other words, the assembly of $\alpha_2\beta_2$ is
negatively cooperative. We have confirmed this finding by gel
chromatography of the preassembled $\alpha\beta_2$ complex (Figure 2), where
the ratio of B-reaction activity in the presence and absence of
an excess of α subunit is plotted against fraction number. Because
the ratio is 50:2:1 for equivalent amounts of $\beta_2:\alpha\beta_2:\alpha_2\beta_2$, it is

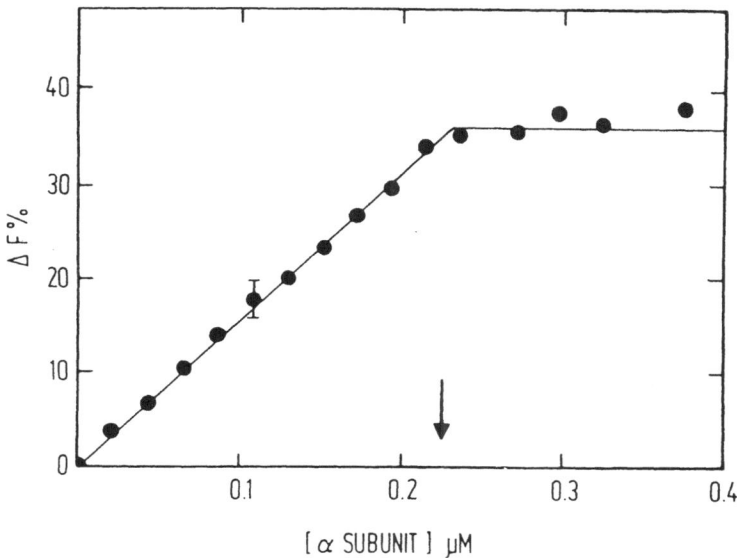

Fig. 1. Fluorimetric titration of β_2 subunit with α subunit. The fluorescence changes ΔF were determined at 25°C with an SLM O.P. 540 model fluorimeter. 0.1 M potassium phosphate buffer pH 7.6 containing 20 μM pyridoxal-P (buffer A). The expected equivalence point ($\alpha/\beta = 1$) is indicated by an arrow. $\left[\text{holo}\beta\right] = 0.23\,\mu M$ ($K_d < 1nM$).

clear that the activity peak contains mainly $\alpha\beta_2$. Although we can offer no plausible physiological rationale for negatively cooperative assembly, this situation is very convenient because the rates of formation of $\alpha\beta_2$ from β_2 (first partial assembly) and $\alpha_2\beta_2$ from $\alpha\beta_2$ (second partial assembly) can be measured separately.

How rapidly does an excess of α subunit bind to the β_2 subunit? After rapid mixing with a stopped-flow apparatus equipped for the detection of fluorescence, the fluorescence increases in two distinct exponential phases and is at equilibrium after 1 second (Fig. 3).

(1) $$\Delta F_t = \Delta F_1^0\, e^{-t/\tau_1} + \Delta F_2^0\, e^{-t/\tau_2}$$

What does this progress curve tell us about the assembly mechanism? Plotting the observed rate constant $1/\tau_1$ against the concentration of excess α subunit shows that $1/\tau_1$ increases indefinitely, which means that the rapid process is second order (Fig. 4A),

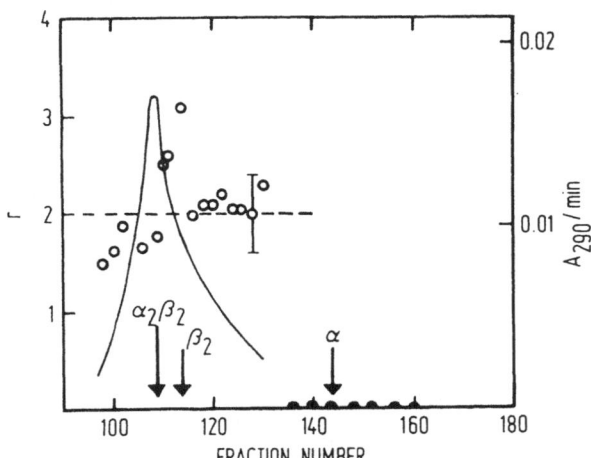

Fig. 2. The holoβ$_2$ subcomplex does not disproportionate in phosphate buffer. Gel permeation chromatography was performed on a 90 x 2.5cm column of Sephacryl S 200 in buffer A at 4°C. 15.6 mg(0.135 μmol) of stoichiometrically preassembled αβ$_2$ subcomplex. Fraction volume 1.9 ml. (●) α subunit activity. (—) αβ$_2$ subcomplex activity. (o) ratio of αβ$_2$ subcomplex activity in presence and absence of excess α subunit. (--) Theoretical ratio for αβ$_2$.

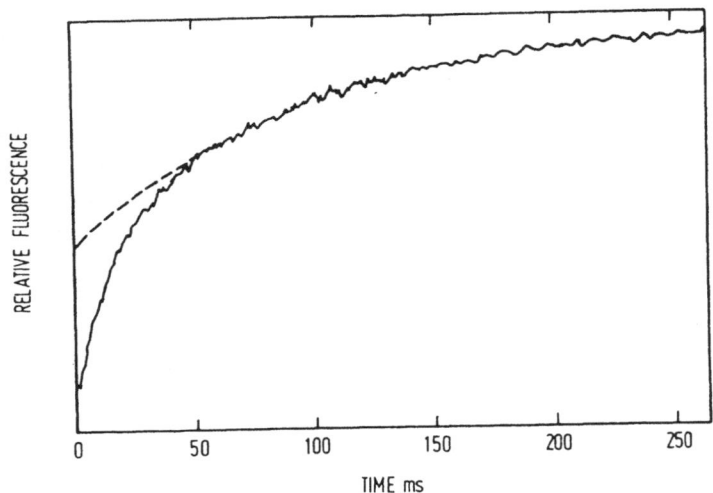

Fig.3. Progress curve showing two exponential processes after rapid mixing of 38 μM α subunit with 4.5 μM holoβ protomer. Buffer A, 25°C. Fluorescence excited at 405 nm, increase of emission recorded above 455 nm. Curve fitting yielded : $1/\tau_1 = 68s^{-1}$, $1/\tau_2 = 10.7s^{-1}$, $\Delta F_1^0 = 0.38$, $\Delta F_2^0 = 0.62$.

corresponding to the formation of an intial $\alpha\beta$ complex. By contrast
$1/\tau_2$ approaches hyperbolically a plateau value at approximately
$16s^{-1}$ with a half-saturation concentration of α subunit ($K_{0.5}$)
at 22 μM (Fig.4B). This behaviour is characteristic of a slow isomerization reaction coupled to a more rapid binding equilibrium. This
two-step mechanism is given by equation (2)

$$(2) \qquad A+B \;\underset{k_D}{\overset{k_R}{\rightleftarrows}}\; AB \;\underset{k_b}{\overset{k_f}{\rightleftarrows}}\; A\overset{*}{B}$$

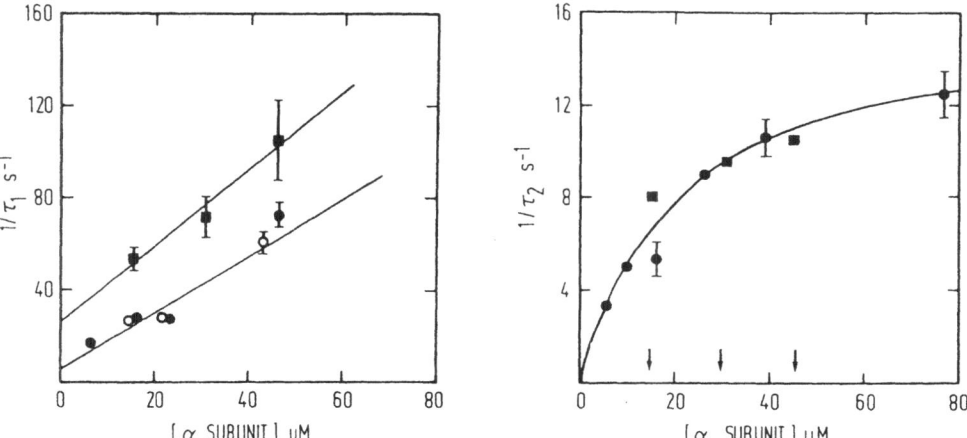

Fig.4. Assembly kinetics: Dependence of $1/\tau_1$ and $1/\tau_2$ on the concentration of excess α subunit. Experimental conditions as under Fig.3.
Error bars indicate standard deviation, straight lines are from
linear regression.(a)Dependence of $1/\tau_1$ on the concentration of excess subunit.($\bullet$) Overall assembly, starting with 3.4 μM holoβ protomers. ($\blacksquare$)Second partial assembly, starting with a preincubated
mixture of 3.5 μM α subunit and 3.5 μM holoβ_2 subunit. (o) Overall
assembly, starting with 3.4 μM holoβ protomers in presence of 0.14
mM indolepropanol phosphate. (b) Dependence of $1/\tau_2$ on concentration of α subunit. ($\bullet$) Overall assembly, starting from 1 μM holoβ
protomers.($\blacksquare$) Second partial assembly, starting from 3.5 μM αholoβ_2
subcomplex. (——) Best fit to equation (4) with k_f = 16.4s^{-1} and
$K_{0.5}$ = 22 μM.

where A is the α subunit, and B, B* are two isomers of the β protomer that differ in the fluorescence quantum yield of bound pyridoxal phosphate. However, these experiments involve the simultaneous
binding of two α subunits to the β$_2$ subunit, in which the first α
subunit is bound more tightly than the second (negative cooperativity). In the following experiments we therefore looked at the binding
of the first and the second α subunit independently.

When α subunit is mixed with an excess of β$_2$ subunit only αβ$_2$
is formed at equilibrium. The progress curves were similar to Fig. 3,
(data not shown) and the concentration dependences of $1/\tau_1$ and
$1/\tau_2$ were qualitatively the same, although quantitatively different.

The rate of the second partial assembly reaction was measured
by mixing the stoichiometrically preassembled αβ$_2$ subcomplex with α
subunit. Again both the progress curves and the concentration dependences resembled the data for overall assembly. They are presented
in Fig. 4A and 4B. Thus it is clear that both the first and the
second αβ protomer are formed via a two-step mechanism. This means
that each of the β protomers isomerizes independently after binding
an α subunit.

Since one of the proteins is always in large excess over the
other, pseudo-first order conditions prevail over the whole concentration range. The dependence of both $1/\tau_1$ and $1/\tau_2$ on concentration
are then given by equations (3) and (4) (Bernasconi, 1976)

(3)
$$1/\tau_1 = k_D + k_R \left[X \right]$$

(4)
$$1/\tau_2 = k_b + k_f \left\{ \frac{\left[X \right]}{\left[X \right] + K_{0.5}} \right\}$$

where the k_n are the rate constants defined in equation (2), $\left[X \right]$ is
the concentration of the subunit present in excess and $k_D/k_R = K_{0.5}$
is the value at which $(1/\tau_2 - k_b)$ is half maximal.

The values of k_D, k_R and k_f obtained for the overall and the
two partial assembly reactions are presented in Table 1. k_b is too
small to be extrapolated from the limited data, and therefore we
cannot yet calculate the overall thermodynamic dissociation constant
$K_D = k_D k_b / k_R k_f$. Nevertheless, it is clear that negative cooperativity is caused at least in part by the different dissociation rate
constants k_D of the initial complexes AB formed, whereas the on-
constant k_R are practically identical. Equations (3) and (4) require

Table 1. Rate and equilibrium constants of the assembly of tryptophan synthase: Analysis of forward rates.

Observed reaction		ligands present	k_R $(\mu M^{-1} s^{-1})$	k_D (s^{-1})	k_f (s^{-1})	k_D/k_R (μM)	$K_{0.5}$ (μM)
excess α + β_2	$\rightarrow$ $\alpha_2\beta_2$	--	1.3	4	16.3	3.1	22
excess β_2 + α	$\rightarrow$ $\alpha\beta_2$	--	1.5	2	6	1.3	4
excess α + $\alpha\beta_2$	$\rightarrow$ $\alpha_2\beta_2$	--	1.6	26	16.3	16.3	22
excess α + β_2	$\rightarrow$ $\alpha_2\beta_2$	Ser	n.d.	n.d.	8	n.d.	10
excess α + β_2	$\rightarrow$ $\alpha_2\beta_2$	IPP	1.4	5	14	3.5	22

The data of Figs. 4A and 4B were analyzed on the basis of equations (3) and (4). IPP: indolepropanol phosphate. n.d. not determined.

Table 2. Rate and equilibrium constants of the assembly of tryptophan synthase:
Analysis of reverse rates.

Observed reaction	ligands present	$10^3 k_b$ (s^{-1})	$10^5 k_b/k_f$	K_D^a (nM)	$\sqrt{K_I K_{II}}$ calc (nM)
α2β2 → β2	--	1.4	8.6	0.27	0.63[b]
αβ2 → β2	--	0.9	15	0.2	--
α2β2 → β2	Ser	0.03	0.38	n.d.	0.0015[c]
α2β2 → β2	IPP	0.2	1.4	0.05	0.063[d]

[a] $K_D = k_D k_b / k_R k_f$. [b] K_D calculated from K_I and K_{II} as described in the text.
[c] 0.63/400. [d] 0.63/10. IPP: indolepropanol phosphate. n.d. not determined.

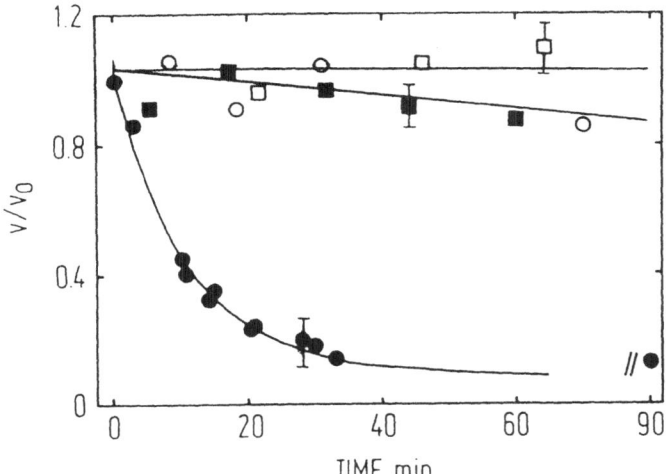

Fig.5. Kinetics of dissociation of the $\alpha_2\beta_2$ complex and the effect of substrates. Dissociation was initiated by adding reduced β_2 subunit and assaying tryptophan synthesis in samples withdrawn after different time intervals. Effect of bound L-serine. ($\bullet$) $\alpha_2$holoβ_2 complex. $k_b = 1.4 \times 10^{-3}$ s^{-1}. ($\blacksquare$) $\alpha_2$holoβ_2 complex plus 2 mM L-serine. $k_b \sim 3 \times 10^{-5}$ s^{-1}. (0, $\square$) Controls with reduced β_2 subunit omitted.

where k_1, k_2 and k_5 correspond to k_R in equation (2), k_{-1}, k_{-3}, k_{-5} to k_D, k_2, k_4 to k_f, and k_{-2}, k_{-4}, k_{-6} to k_b. The values of some of these rate constants have been measured directly (Cf. Tables 1 and 2). The others can be estimated from the rate constants observed for the overall assembly. Since the simultaneous binding of two α subunits can be described by the same formalism as the two partial assembly reactions separately (equations (2) to (4)), it is likely that similar processes have similar rate constants and fluorescence amplitudes. Therefore the observed rate processes must be degenerate. The detailed arguments are given elsewhere (Lane and Kirschner, 1984), and the estimated rate and microscopic equilibrium constants are given in Table 3.

The microscopic equilibrium constants give an estimate of the degree of negative cooperativity. At equilibrium, the assembly involves only the species of equation (7)

Table 3. Rate and equilibrium constants of the assembly of the $\alpha_2\beta_2$ complex of tryptophan synthase.

| constant | binding steps | | constant | isomerization steps | |
	dimension	value		dimension	value
k_1	$\mu M^{-1} s^{-1}$	1.5	k_2	s^{-1}	6
k_{-1}	s^{-1}	2	k_{-2}	s^{-1}	0.9×10^{-3}
K_1	μM	1.3	K_2	--	1.5×10^{-4}
k_3	$\mu M^{-1} s^{-1}$	n.d.	k_4	s^{-1}	(16)
k_{-3}	s^{-1}	n.d.	R_{-4}	s^{-1}	(1.5×10^{-3})
K_3	μM	(26)	K_4	--	(0.9×10^{-4})
k_5	$\mu M^{-1} s^{-1}$	1.6	k_6	s^{-1}	16.3
k_{-5}	s^{-1}	26	k_{-6}	s^{-1}	1.9×10^{-3}
K_5	μM	16.3	K_6	--	1.2×10^{-4}

The data of Tables 1 and 2 were interpreted on the basis of equation (6) as described in the text. Values in parentheses are estimates. n.d. not determined.

that $K_{0.5} = k_D/k_R$. This prediction is fulfilled well for the second partial assembly, but not so well for the first partial assembly. Whereas the latter discrepancy is probably due to the poor signal-to-noise ratio, the even larger discrepancy between $K_{0.5}$ and k_D/k_R for the overall assembly must be due to a superposition of the first and second partial assembly processes.

To measure the elusive reverse isomerization constant k_b we followed the approach of Creighton and Yanofsky (1966) by mixing either $\alpha\beta_2$ or $\alpha_2\beta_2$ with an excess of an inactive derivative of the β_2-subunit. This protein traps released α subunit, and is obtained by reduction of the Lys_{86}-pyridoxal phosphate azomethine bond with $NaBH_4$ (Wilson and Crawford, 1965). Inspection of Figs. 4A and 4B shows that $k_b << k_D$, and therefore the rate of complete dissociation of $\alpha_2\beta_2$ to β_2 is limited by k_D, as given by equation (5)

$$(5) \qquad AB \xrightarrow[k_b]{\ast} (AB) \xrightarrow[k_D]{} (A+B) \xrightarrow[k_{R'}]{B'} AB' + B$$

where the rate constants are as defined for equation (2) and B' is the reduced β protomer. We measure k_b as the first-order rate constant by which enzyme activity in the B reaction is reduced 50-fold as AB is converted to B (Fig. 5).

The values of k_b measured for both $\alpha\beta_2$ and $\alpha_2\beta_2$ are presented in Table 2, together with the calculated isomerization ($K=k_b/k_f$) and overall dissociation ($K_D = k_D k_b/k_R k_f$) equilibrium constants. It is apparent that K_D of $\alpha_2\beta_2$ is smaller than 1 nM (cf. Fig.2) and that the slow isomerization reaction contributes strongly to the high stability of the $\alpha\beta$ interaction.

By combining the kinetic data on the first and the second partial assembly reactions we propose the following mechanism

$$(6)$$

(7)
$$\alpha + \beta_2 \xrightleftharpoons{K_I} \alpha\overset{*}{\beta}\beta \xrightleftharpoons{K_{II}} \alpha_2\overset{*}{\beta}_2$$

where $K_I = K_1 K_2 / (1 + K_2) \sim K_1 K_2 = 0.20$ nM and $K_{II} \sim K_5 K_6 = 1.96$ nM. The calculated Hill coefficient (Levitzki, 1975; Lane, 1982) is $n_H = 2 / (1 + \sqrt{K_I K_{II}}) = 0.55$, i.e. smaller than 1, as expected for negative cooperativity. At 50% saturation the concentration of free α subunit $[\alpha_{0.5}] = \sqrt{K_I K_{II}} = 0.63$ nM (Levitzki, 1975) can be used to estimate the equilibrium distribution between $\beta_2 : \alpha\beta\beta : \alpha_2\beta_2$ at half saturation. The calculated ratios are 0.14 : 0.72 : 0.14, which agrees qualitatively with the enhancement effects observed upon gel filtration of the pre-assembled $\alpha\beta_2$ complex (Cf. Fig. 2). Thus the kinetic data are consistent qualitatively and semiquantitatively with the equilibrium data.

At which stage in the assembly mechanism are the α and β protomers enhanced in their enzymic activities and affinities for substrates ? We have attempted to answer this question for each subunit separately by exploiting the spectral changes that accompany the binding of specific ligands.

In the case of the β protomer, L-serine is a suitable ligand. It is the leading substrate in the B-reaction and causes large changes in the fluorescence of pyridoxal phosphate at the active site. Moreover the β_2 subunit forms only the external aldimine of L-serine, Q, with its α carbon proton still in place (Lane and Kirschner, 1983a), whereas the mature $\alpha_2\beta_2$ complex, $\overset{*}{A}B$ converts the external aldimine in a sequence of two additional steps to the external aldimine of α aminocrylate, $\overset{*}{A}P$, (Lane and Kirschner, 1983b).

(8)
$$B + Ser \rightleftharpoons Q$$

(9)
$$A\overset{*}{B} + Ser \rightleftharpoons \overset{*}{A}Q \rightleftharpoons \overset{*}{A}F \rightleftharpoons \overset{*}{A}P$$

L-serine has lost its α carbon proton in $\overset{*}{A}F$. Moreover, $\overset{*}{A}Q$ is formed very rapidly from $A\overset{*}{B}$, whereas Q is formed only slowly from B (Lane and Kirschner, 1983a, 1983b).

The results of rapid mixing experiments (not shown here) in which the β_2 subunit was either preincubated with L-serine and then mixed with excess α subunit (i.e. A+Q → AQ → → → $\overset{*}{A}P$), or was mixed with a solution of α subunit that also contained L-serine, indicate the following : (i) The initial complex AB binds L-serine

more rapidly than does the β_2 subunit; (ii) the initial complex
also catalyzes rapidly the deprotonation of AQ, and (iii) the dehy-
dratation reaction as accelerated mainly in the slow isomerization
phase. In other words, the β protomer is already partially activated before the isomerization has occurred. This correlates with
the interesting but still unexplained activating effect of high
concentrations of ammonium ions (Crawford and Ito, 1964; Miles and
McPhie, 1974).

Bound L-serine also slows down strongly the reverse isomerization constant k_b (Cf. Fig. 5 and Table 2), which accounts partially
for the 400-fold enhancing effect of bound L-serine (i.e., the
external aldimine of α aminoacrylate) on the $\alpha\beta$ interaction
(Lane and Kirschner, 1983a and 1983b). More work is required to sort
out the effects of L-serine and its different derivatives on the
rates and equilibria of the initial binding process $A + B \rightleftharpoons AB$
(Cf. Equation 2).

What about the α subunit ? In this case the substrate analogue
indolepropanol phosphate is a suitable chromophoric ligand. It
binds to the α subunit , but 10-fold more strongly to the $\alpha_2\beta_2$
complex (Kirschner and Wiskocil, 1975). Figure 4A shows that pre-
incubating the α subunit with a saturating concentration of indolepropanol phosphate does not change the kinetics of the rapid binding
step. The value of k_f, is also practically the same, but k_b is
about 10-fold smaller than in absence of the ligand (Cf. Table 2).
This effect accounts almost completely for the overall 10-fold enhancement of the $\alpha\beta$ interaction by bound indolepropanol phosphate,
indicating that the high affinity for the ligand is only gained
during the slow isomerization step, and that indolepropanol phosphate stabilizes a high affinity state of the α protomer, but does
not induce it. This interpretation was confirmed directly (Lane and
Kirschner, 1984) by measuring the rate of binding of indolepropanol phosphate during the assembly (data not shown). These observations mean that the slow isomerization process involves a synchro-
nous rearrangement of both the α and β protomer.

DISCUSSION

Our work has shown conclusively that the assembly of tryptophan
synthase occurs in two distinct steps. Each β protomer behaves
essentially independently, as far as the rate and equilibrium
constants of the isomerization processes are concerned (Cf. Table 3).
Because negative cooperativity is expressed almost exclusively
by the larger off constant of the second initial $\alpha\beta$ complex
($k_{-5} > k_{-1}$), it is likely that steric hindrance between the α subunits is the cause. If this is correct, the slow isomerization process must involve a mutual reorientation of the $\alpha\beta$ protomers from
an initial configuration, in which the α subunits touch each other,

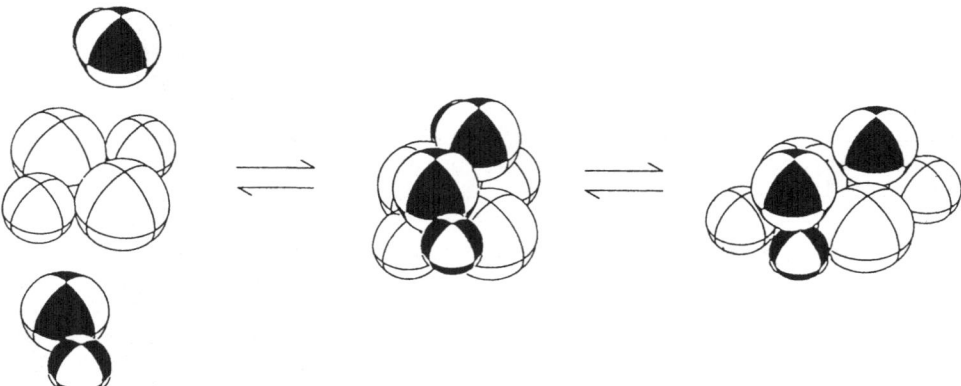

Fig. 6. Quaternary changes during the assembly of tryptophan synthase : a plausible model. The models of the α and β subunits and
the α2β2 complex were taken from Lane and Kirschner (1983c). The
intermdiate structure was obtained by packing α subunits on top
of the β2 subunit so that they touch. It does not account for the
data as well as the final structure, in which the structure of the
β2 subunit is altered.

to one, in which the strain is relieved. Low resolution models of
the subunits and the complex of tryptophan synthase (Wilhelm et al,
1982; Lane and Kirschner, 1983) support this hypothesis (Fig. 6).
The best models for the $\alpha_2\beta_2$ complex cannot be generated merely by
packing the best models for the separated subunits. Moreover,
recent results from neutron small-angle scattering studies on partially deuterium-labelled $\alpha_2\beta_2$ complex (K. Ibel, R.May, A.N. Lane
and K. Kirschner, unpublished data) have shown that the α subunits
are separated by at least 20 Å, confirming an important aspect of
the model of the $\alpha_2\beta_2$ complex.

The forward isomerization rate constants k_f (Table 1) are
at least four order of magnitude larger than the slow "reshuffling"
rate constants observed in the refolding and reassociation of many
denatured oligomeric enzymes (Jaenicke, 1982). Since both the α
and β protomers consist of autonomously folding and strongly interacting structural domains (Miles, 1979; Zetina and Goldberg, 1982),
it is possible that mutual rearrangements of the four different domains of tryptophan synthase are responsible for the slow isomerization (Janin and Wodak, 1983). This hypothesis is particularly
attractive, since indirect evidence (Miles, 1966) indicates that the
active sites of both α and β lie in the crevices between the cognate
domains.

ACKNOWLEDGEMENTS

This study was supported by grant 3.744.80 from the Swiss National Science Foundation.

REFERENCES

Adachi, O., Kohn, L.D. and Miles, E.W. (1974) J. Biol. Chem. 249, 7756-7763.

Bernasconi, C. (1976) "Relaxation kinetics", Academic Press, New-York.

Bothwell, M.A. and Schachman, H.K. (1980) J. Biol. Chem. 255, 1971-1977.

Crawford, I.P. and Ito, J. (1964) Proc. Natl. Acad. Sci. USA 51, 390-397.

Creighton, T.E. and Yanofsky, C. (1966) J. Biol. Chem. 241, 980-990.

Friedman, F.K. and Beychok, S. (1979) Ann. Rev. Biochem. 48 , 217-250.

Jaenicke, R. (1982) Biophys.Struct. Mech. 8 , 231 -256.

Janin, J. and Wodak, S.J. (1983) Progr. Biophys. Molec. Biol. 42, 21-78.

Kirschner, K., Wiskocil, R.L., Foehn, M. and Rezeau, L. (1975) Eur. J. Biochem. 60, 513-523.

Lane, A.N.(1982) J. Theor. Biol. 99, 491-508.

Lane, A.N. and Kirschner, K. (1983a) Eur. J. Biochem. 129,561-570.

Lane, A.N. and Kirschner, K. (1983b) Eur. J. Biochem. 129, 571-582.

Lane, A.N. and Kirschner, K. (1983c) Eur. J. Biochem. 129, 675-684.

Lane, A.N. and Kirschner, K. (1984) EMBO J. 4, 279-287.

Levitzki., A. (1975) in "Subunits Enzymes, Biochemistry and Function" Ebner, K.E. ed. pp. 1-41., Marcel Dekker, New-York.

McDonald, M.K. (1981) J. Biol. Chem. 256,6487-6490.

Miles, E.W. (1970) J. Biol. Chem. 245, 6016-6025.

Miles, E.W. (1979) Advanc. Enzymol. Relat. Areas Mol. Biol. 49, 127-186.

Miles, E.W. and McPhie, P. (1974) J. Biol. Chem. 249, 2852-2857.

Mosteller, R.D., Goldstein, R.V. and Nishimoto, K.R. (1977) J. Biol. Chem. 252 , 4527-4532.

Tainski, M. and Edelstein, S.J. (1973) J. Mol. Biol. 75 , 735-739.

Welch, G.R. (1977) Progr. Biophys. Mol. Biol. 32, 103-191.

Wilhelm, P., Pilz, I., Lane, A.N. and Kirschner, K. (1982) Eur. J. Biochem. 129 , 51-56.

Wilson, D.A. and Crawford, I.P. (1965) J. Biol. Chem. 240, 4801-4808.

Yanofsky, C. and Crawford, I.P. (1972) in "The Enzymes" , P.D. Boyer ed. 3rd edn. Vol. 7, pp. 1-31, Academic Press, New-York.

York, S.S. (1972) Biochemistry 11 , 2733-2740.

Zetina, C.R. and Goldberg, M.E. (1982) J. Mol. Biol. 157, 133-148.

SECTION IV - DYNAMICS OF ENZYME REACTIONS
IN HETEROGENEOUS MEDIA

ELECTRIC REPULSION EFFECTS AND THE DYNAMIC BEHAVIOUR OF ENZYMES

EMBEDDED IN BIOLOGICAL POLYELECTROLYTES

J. Ricard, G. Noat and M. Crasnier

Centre de Biochimie et de Biologie Moléculaire du CNRS
31 Ch. J. Aiguier, BP 71, 13402 Marseille Cedex 9

INTRODUCTION

The dynamic behaviour of an enzyme within an insoluble polyelectrolyte, such as a membrane or a cell wall, may be quite different from the one it should display if it were in free solution.
Three reasons at least may explain this change of behaviour :
-the enzyme conformation may have been changed upon binding to the
polyelectrolyte;
-the substrate of the enzyme may be submitted to an electrostatic
partition effect;
-diffusional resistances within the polyelectrolyte matrix may be
such that a coupling occurs between enzyme reaction rate and diffusion of substrate and product.

Although the first effect is trivial, the second and the third
are rather subtle and may generate complex enzyme behaviour, which
may play an important role in the control of metabolic networks.
When applied to an enzyme located at the outer surface of a living
cell, the above ideas may be particularly important. They provide
a rationale which allows understanding the physical basis of elementary perception of signals from the external milieu and their
conduction. Since isolated plant cells in sterile culture are
surrounded by a polyanionic primary cell wall, which is the physical support of several hydrolytic enzymes, this system is certainly
well suited for studying the dynamic behaviour of enzymes in heterogeneous charged media.

The aim of this report is to study how the interplay between
enzyme reaction, electric partition and diffusional resistances of

substrate and product, may create unexpected enzyme dynamic behaviour at the surface of living cells. Moreover it is important to
know whether these dynamic effects may take part in the elementary
perception and conduction of signals from the external milieu.

ELECTROSTATIC PARTITION OF SUBSTRATE AND KINETICS OF A BOUND ENZYME

If an enzyme is homogeneously distributed within an insoluble
polyelectrolyte matrix and if the substrate of the enzyme is
itself an ion, its concentration inside the matrix is usually different from its concentration outside that matrix. If no diffusional
resistance occurs one may define an electrostatic partition coefficient Π from the ratio of the concentrations of the various ions inside and outside the matrix. If there exists inside and outside the
matrix different ions of different valency, B^- , B^{2-}, ..., B^{z-} and
different monovalent cations A^+, the electrostatic partition coefficient is defined as (see Figure 1)

$$\frac{\Sigma[B_o^-]}{\Sigma[B_i^-]} = \frac{\Sigma[B_o^{2-}]^{1/2}}{\Sigma[B_i^{2-}]^{1/2}} = \cdots = \frac{\Sigma[B_o^{z-}]^{1/z}}{\Sigma[B_i^{z-}]^{1/z}} = \frac{\Sigma[A_i^+]}{\Sigma[A_o^+]} = \Pi$$

(1)

$$\Pi = \exp(F\Delta\bar{\psi}_D / RT)$$

The subscripts refer to the inside and the outside of the
matrix, $\Delta\bar{\psi}_D$ is the electrostatic potential difference between
the bulk phase and the Donnan phase , namely $\Delta\bar{\psi}_D = \bar{\psi}_D^o - \bar{\psi}_D^i$. F is the
Faraday, R and T have their usual significance. If Δ^- is the "densi-
ty" or the "concentration" of the fixed charges inside the matrix,
the electro-neutrality equation allows to express the partition
coefficient as a function of ionic concentrations in the bulk phase.
One finds

$$\Pi^{z+1} \pm \frac{\Delta^\pm}{\Sigma[B_o^-]+\ldots+z\Sigma[B_o^{z-}]} \Pi^z - \frac{\Sigma[B_o^-]}{\Sigma[B_o^-]+\ldots+z\Sigma[B_o^{z-}]} \Pi^{z-1}$$

(2)

$$-\frac{2\Sigma[B_o^{2-}]}{\Sigma[B_o^-]+\ldots+z\Sigma[B_o^{z-}]} \Pi^{z-2} -\ldots- \frac{z\Sigma[B_o^{z-}]}{\Sigma[B_o^-]+\ldots+z\Sigma[B_o^{z-}]} = 0$$

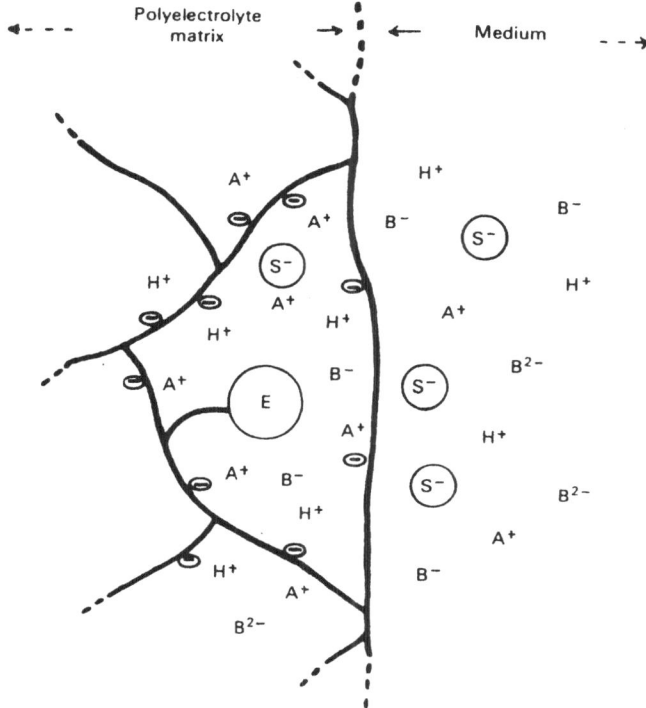

Fig. 1. Electrostatic partition of an anionic substrate by a poly-
anionic matrix (From Riçard et al, 1981).
The enzyme E is homogeneously distributed in the matrix. S⁻ is a
monovalent substrate, B⁻ a monovalent anion, B^{2-} a di-anion and
A^+ a cation.

    If the enzyme reaction follows Michaelis-Menton kinetics
and if the substrate is a monovalent anion, one has

$$(3) \qquad v = \frac{V_{max}\,[S_i^-]}{K_m + [S_i^-]}$$

which may be rewritten as

$$(4) \qquad v = \frac{V_{max}\,[S_o^-]}{\Pi K_m + [S_o^-]}$$

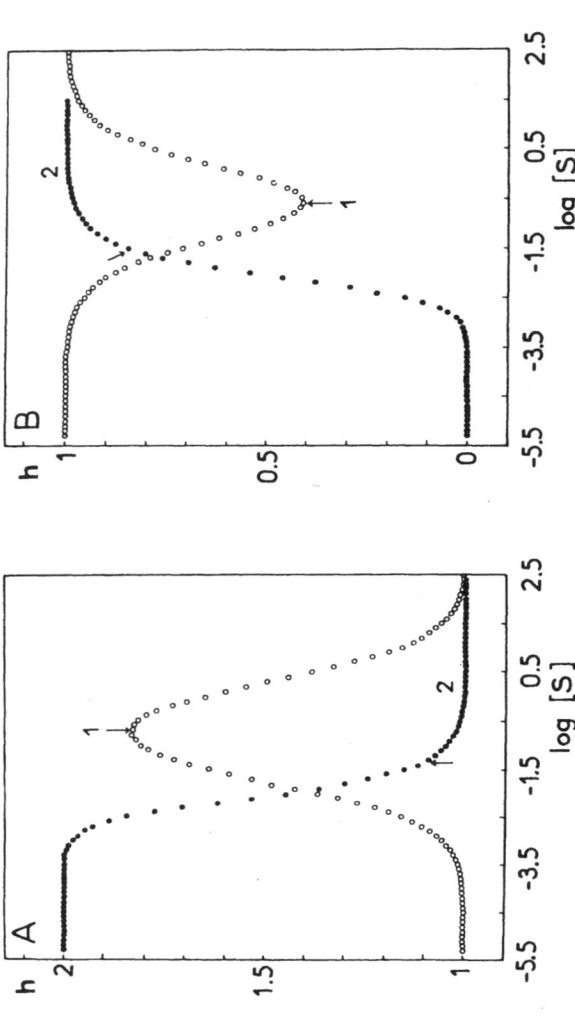

Fig.2. Comparative variation of the Hill coefficient (h) predicted by the Adair equation and by electrostatic repulsion effects.

A) Positive co-operativity. Curve 1 corresponds to the Adair equation for a dimeric enzyme. h becomes maximum at half-saturation of the enzyme by the substrate (arrow). Curve 2 shows the variation of the Hill coefficient due to electrostatic repulsion of the substrate.

B) Negative co-operativity. Curve 1 corresponds again to the Adair equation for a dimer. h becomes minimum at half-saturation of the enzyme by the substrate (arrow). Curve 2 shows the variation of the Hill coefficient due to electrostatic attraction of the substrate.

In either case the enzyme is assumed to be homogeneously distributed in the matrix. For further details see Ricard et al, 1981.

Clearly, as substrate concentration is varied, the partition coefficient Π varies as well. Therefore a plot of v <u>versus</u> $[S_o^-]$ may well not be hyperbolic. Engasser and Horvath (1975) as well as Ricard et al (1981) have shown that if the fixed charges of the Donnan phase tend to repel the substrate, an apparent positive co-operativity is observed. Alternately if the fixed charges of the matrix tend to attract the substrate, an apparent negative co-operativity may be detected. Obviously if the partition coefficient Π does not vary significantly as the substrate concentration is varied the bound enzyme still displays a classical Micahelis-Menten behaviour but its Km is modified by a factor of Π.

If Π varies as a function of $[S]$, the variation of the Hill coefficient is shown in Figure 2 for both substrate attraction and repulsion.

By contrast with the classical co-operativity predicted by the Adair equation, which is maximum in the intermediate range of concentrations, the apparent co-operativity of the bound enzyme, generated by electrostatic repulsion or attraction of the substrate, is maximum or minimum at low substrate concentration (Figure 2).

It is worth noting that this apparent co-operativity is not the property of the enzyme but of the system made up with the association of the enzyme with the insoluble polyelectrolyte matrix. An obvious important property of this system is that upon increasing the ionic strength, the value of the partition coefficient approaches unity and the bound enzyme behaves as if it were in free solution.

We have considered thus far the case where the enzyme was homogeneously distributed within the polyelectrolyte matrix. It may well occur in fact that a population of the enzyme molecules is located at the interface between the bulk phase and the matrix and is not submitted, or poorly submitted, to the electrostatic effects whereas another population is deeply buried in the matrix and therefore strongly dependent on electrostatic effects (Figure 3). If the substrate is a monovalent anion and if the matrix is poly-anionic, the relevant enzyme reaction rate is now

(5) $$v = \frac{V_{max1}\,[S_o^-]}{K_m + [S_o^-]} + \frac{V_{max2}\,[S_o^-]}{\Pi\,K_m + [S_o^-]}$$

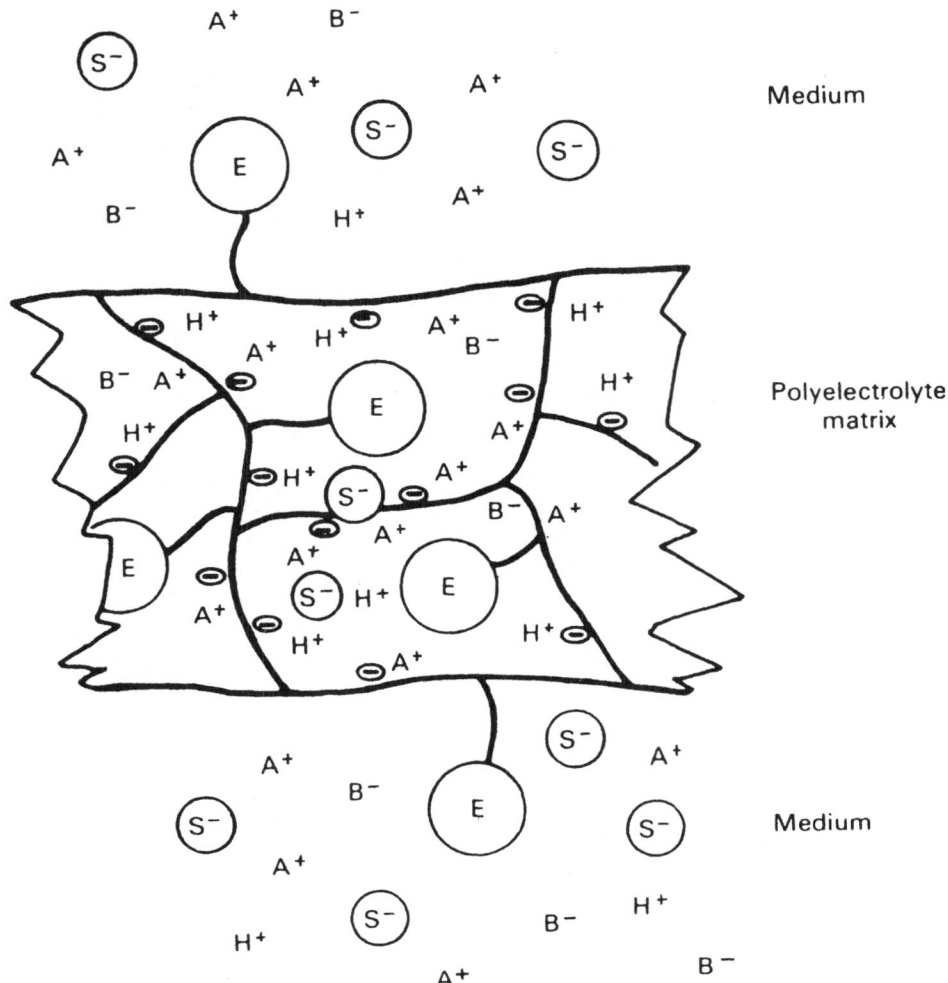

Fig.3. Electrostatic partition of an anionic substrate by a poly-anionic matrix (from Ricard et al, 1981). The enzyme E is heterogeneously distributed in the matrix. S^- is the monovalent substrate, B^- a monovalent anion, B^{2-} a divalent anion and A^+ a cation.

where V_{max1} and V_{max2} represent the maximum reaction rates of the two enzyme populations. If the partition coefficient does not vary significantly as the substrate concentration is varied, the overall system exhibits an apparent negative co-operativity, for the situation is then equivalent to a reaction conditioned by two different enzymes acting on the same substrate (Dixon and Webb, 1979). If the partition coefficient varies as the substrate concentration is varied, the kinetic behaviour may even be more complex and show mixed positive and negative co-operativity (Figure 4).

The kinetic study of an acid phosphatase bound to sycamore cell walls allows to check the validity of the above model. An acid phosphatase may be isolated and purified to homogeneity from sycamore cell walls. It is a glycoprotein of 100 000 molecular weight.

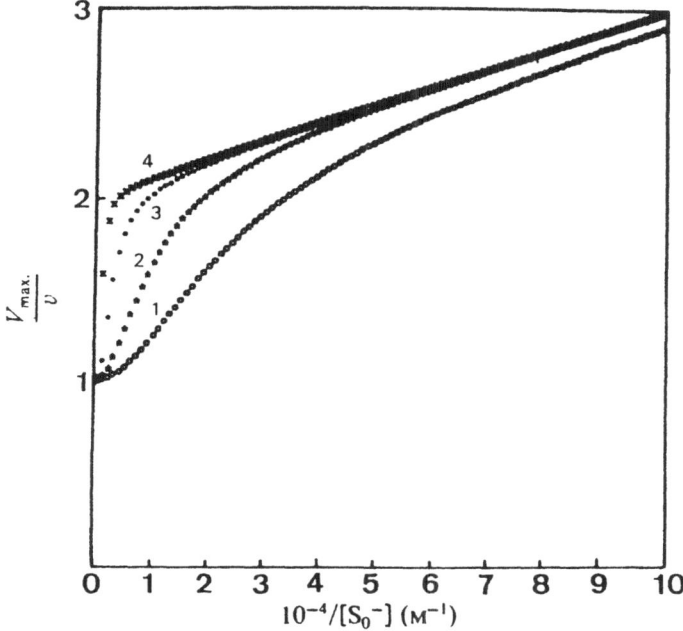

Fig. 4. Apparent mixed positive and negative co-operativity due to electrostatic repulsion of the substrate (from Ricard et al, 1981). The enzyme is assumed to be heterogeneously distributed within the matrix and the partition coefficient varies as the substrate concentration is varied. Electrostatic repulsion effects increase from curve 1 to 4.

When following hydrolysis of p-nitrophenyl phosphate, the enzyme in free solution displays classical Michaelis-Menten kinetics. A similar result is obtained with the enzyme bound to cell wall fragments or to unbroken isolated cells (Figure 5).

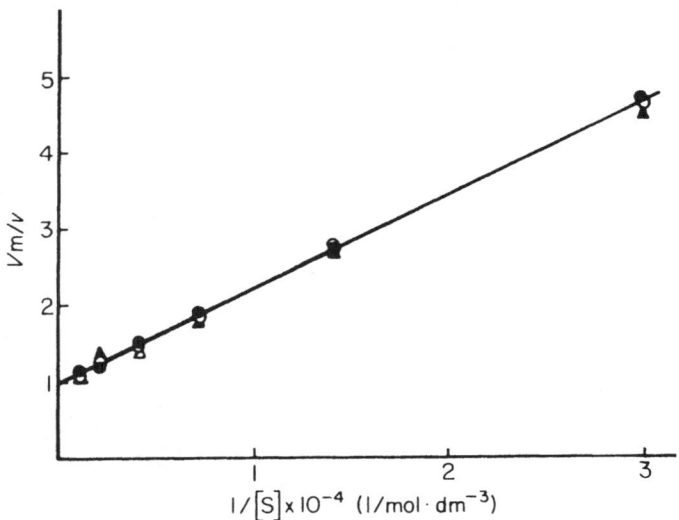

Fig.5. Identity of kinetic behaviour of the soluble and the bound
acid phosphatase. p-nitrophenylphosphate hydrolysis is determined
with either intact cells (-o-) , a suspension of cell wall fragments
(-•-) or with a purified enzyme in free solution(-▲-) (from Crasnier
et al , 1980a).

This identity of kinetic behaviour however, is obtained at
high ionic strength only. At low ionic strength and at constant
substrate concentration, the reaction rate increases as the ionic
strength is increased (Figure 6).

These results strongly suggest that electrostatic repulsion
of p-nitrophenyl phosphate is exerted by the fixed negative charges
of the cell wall .

The existence of these fixed negative charges may be easily
detected experimentally by titration (Figure 7A). Titration of
cell wall fragments at low ionic strength shows the existence of
an apparent buffering effect at low ionic strength. This effect
does not exist when ionic strength is increased. The existence of
this buffering effect implies that upon titration, protons are
released from the cell walls . The existence of high proton concentrations in the cell wall becomes obvious upon raising the ionic
strength of a suspension of cell wall fragments. Under these conditions an outflux of protons is observed and may be titrated (Figure
7B).

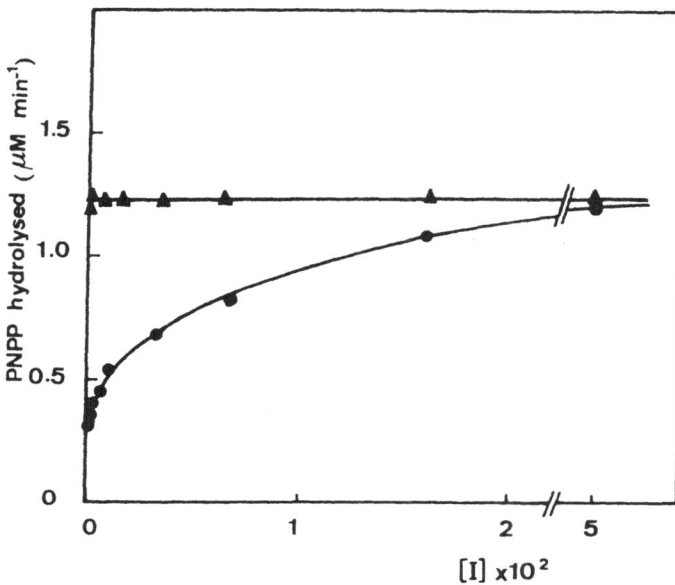

Fig.6. Ionic strength dependence of cell wall acid phosphatase activity. The enzyme activity of the soluble enzyme (-▲-) and of the bound enzyme (-●-) is measured at constant substrate concentration and at varied ionic strength. The ionic strength is adjusted with NaCl (from Noat et al, 1980 b).

Although the bound acid phosphatase exhibits a Michaelis-Menten behaviour at high ionic strength, it exhibits a negative kinetic co-operativity at low ionic strength (Figure 8). This is clearly understandable on assuming that a population of enzyme molecules is located at the interface between the bulk phase and the Donnan phase, whereas other enzyme molecules are buried in the matrix and submitted to electrostatic partition effects of substrate. As the ionic strength is raised, the bound enzyme becomes activated and tends to follow Michaelis-Menten kinetics (Noat et al, 1980).

ELECTRIC PARTITION UNDER NON-EQUILIBRIUM CONDITIONS

So far we have considered that diffusional resistances of substrate at the surface of cell walls, or in cell wall fragments, is negligible. The result that, at high ionic strength, reaction rate of the bound enzyme is similar to that of the free enzyme clearly shows that this assumption is supported by experiment.

It may occur however that, in an unstirred medium, diffusional resistances of substrate and product exist together with electric partition and enzyme reaction. Under these non-equilibrium condi-

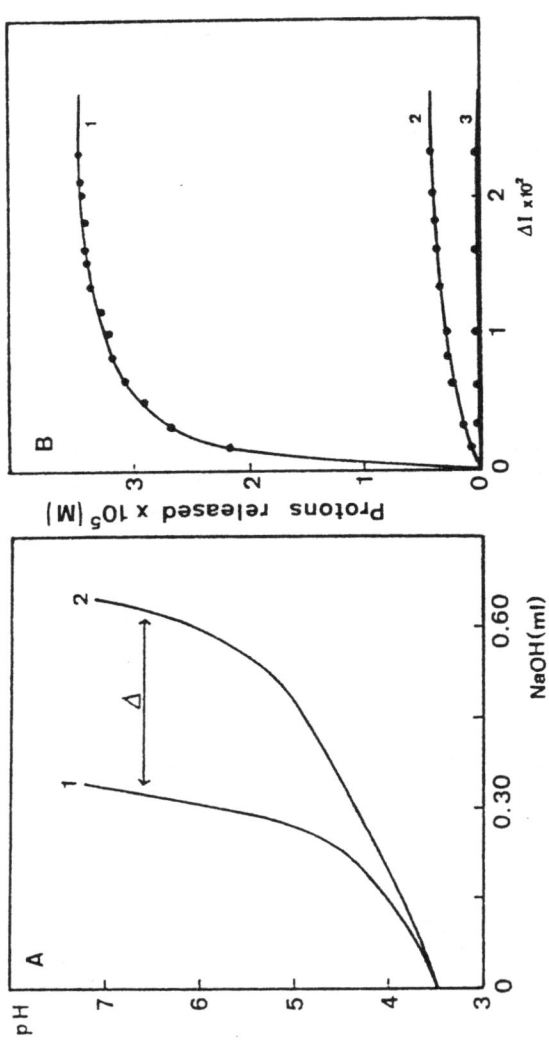

Fig.7. Polyanionic properties of plant cell wall fragments.
A/ Ionic strength effect on the titration curve of cell wall suspensions. Titration was effected in the presence of NaCl 0.1M (curve 1) or without salt (curve 2). Before starting the titration the pH was adjusted to 3.5. Titration was then carried out with NaOH. Δ reflects the extent of an electrostatic effect in the cell walls.
B/ Proton extrusion from the cell wall fragments, generated by raising the ionic strength of the medium. Curves 1 and 2 are obtained without salt (curve 1) or with 0.01 M NaCl (curve 2) at the starting time. Curve 3 is a blank containing no cell wall fragments.

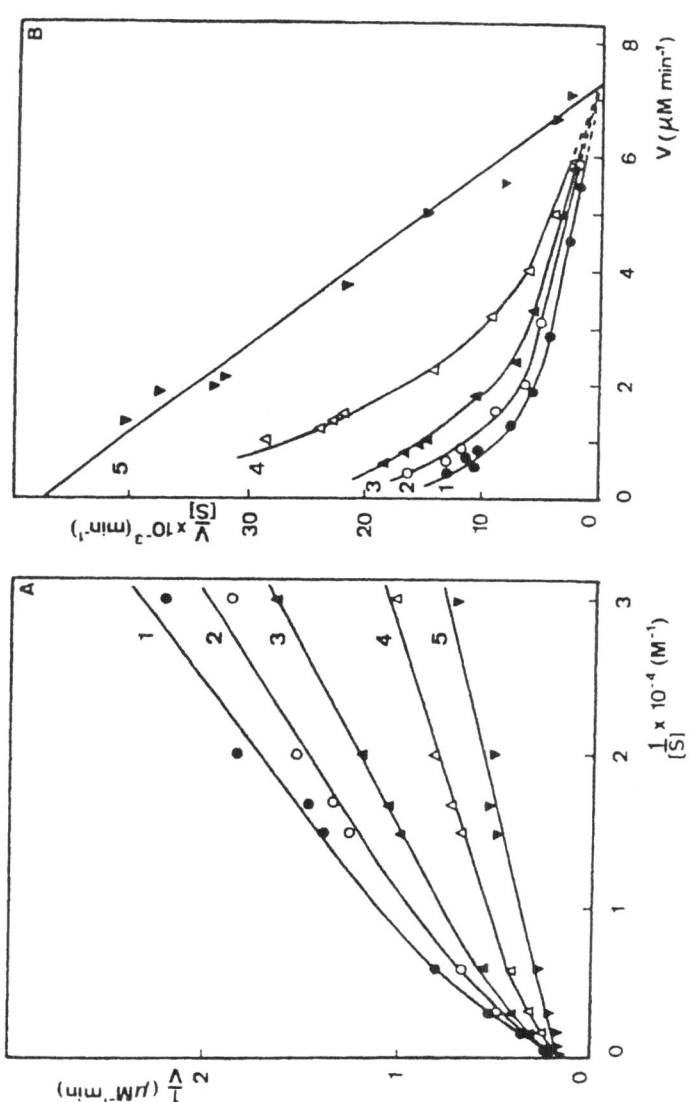

Fig. 8. Apparent negative co-operativity of bound acid phosphatase at low ionic strength.
A) Lineweaver-Burk plots   B) Eadie plots
In either panel curves 1 to 5 are obtained by increasing concentrations of NaCl. Data are from Ricard et al (1981).

tions one may wonder whether the concepts of electric partition coefficient and Donnan potential still apply.

Let there be the simple one-substrate , two-product reaction

$$E \xrightarrow[k_{-1}]{k_1[S]} (ES,\ EPQ) \xrightarrow[P]{k_2} EQ \xrightarrow[k_{-3}\ [Q]]{k_3} E$$

where S, P and Q are the substrate and the two products. In this scheme, it is postulated that the first step of product release is nearly irreversible, for the appearance of P is determined under initial steady state conditions. This situation is precisely the one known to occur for most hydrolytic enzymes and in particular for acid phosphatase of plant cell walls. Clearly this reaction scheme cannot predict any co-operativity. In order to set up a tractable model of the coupling between diffusion, electric partition effects and reaction, a layer of enzyme molecules is assumed to be located on an impermeable surface and embedded in a thin layer of a polyanionic matrix (Figure 9.). Under steady state conditions the flux of substrate ion which reaches the surface may be shown to be of the form

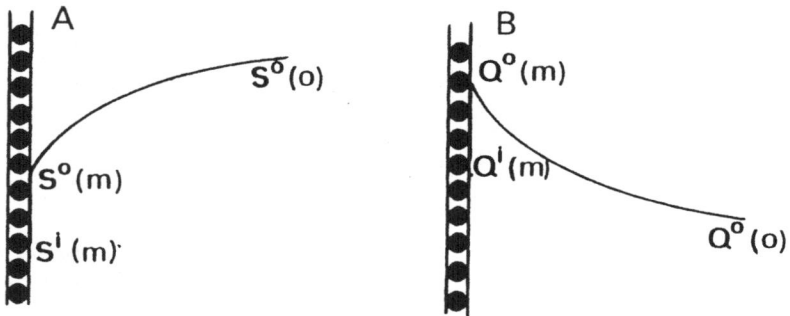

Fig. 9. Coupling between substrate and product diffusion , electric repulsion and enzyme reaction.
A / Coupling between substrate diffusion, electric partition and enzyme reaction.
B / Coupling between product diffusion , electric partition and enzyme reaction.

(6)
$$J_s = \tilde{h}_d \{\tilde{s}^o(o) - s^o(m)\}$$

where

(7)
$$\tilde{h}_d = \frac{zF\Delta\psi^o}{RT \{\exp \frac{zF\Delta\psi^o}{RT} - 1\}}$$

and

(8)
$$\tilde{s}^o(o) = s^o(o) \exp \left(\frac{z F \Delta\psi^o}{RT}\right)$$

$s^o(o)$ is the substrate concentration in the bulk phase, $s^o(m)$ the corresponding external concentration in the immediate vicinity of the surface, h_d the transport constant, $\Delta\psi^o$ the diffusion potential in the interval $[o,m]$, between the bulk phase and the surface, z the valency of substrate, F the Faraday, R and T have their usual significance. The flux of product which leaves the surface may be expressed in a way similar to that shown in equation (6).

The substrate transport from the distance 0 to the layer of enzyme molecules in the matrix at the distance m, may be expressed by the following kinetic diagram

$$\text{Input} \xrightarrow{\quad\quad} s^o(o) \underset{\tilde{h}_d}{\overset{\tilde{h}_d}{\rightleftharpoons}} s^o(m) \underset{\tilde{h}_d\beta}{\overset{\tilde{h}_d\alpha}{\rightleftharpoons}} s^i(m) \xrightarrow{\quad\quad} \text{Output}$$

where α and β are electric coefficients which may be found to be (Hill and Chen, 1970, 1971, Hill, 1977).

(9)
$$\alpha = \exp (-F\Delta\psi/2RT)$$
$$\beta = \exp (F\Delta\psi/2RT)$$

$\Delta\psi$ is the potential difference between the external solution at the surface of the matrix and the inside of that matrix. Obviously $\Delta\psi$ may be identitified to a Donnan potential $\Delta\bar{\psi}_D$ in the vicinity of the surface, if the step

$$S^o(m) \xrightleftharpoons[\tilde{h}_d\beta]{\tilde{h}_d\alpha} S^i(m)$$

is in a pseudo-equilibrium state whereas the whole system is in steady state. It may be shown that this condition is satisfied if

$$(10) \qquad\qquad \exp(F\Delta\psi/2RT) \gg 1 + \exp(-F\Delta\psi/2RT)$$

This inequality holds if $\Delta\psi$ is large. Then a Donnan potential may be defined at the surface of the polyelectrolyte and an electric partition coefficient Π may thus be defined under non-equilibrium conditions.

COUPLING BETWEEN DIFFUSION, REACTION AND ELECTRIC PARTITION OF SUBSTRATE AND PRODUCT

Since the concept of electric partition coefficient may be applied under non-equilibrium conditions at the surface of the matrix, the flux of substrate which reaches the layer of enzyme molecules may be expressed as

$$(11) \qquad\qquad J_S = \tilde{h}_d \{ \tilde{S}{}^o(o) - \Pi^z S^i(m) \}$$

Alternately the flux of product which leaves these enzyme molecules is

$$(12) \qquad\qquad J_Q = \tilde{h}'_d \{ \Pi^{\lambda z} Q^i(m) - \tilde{Q}{}^o(o) \}$$

where λ is such that the product λz is the valency of the product and $\tilde{h}'_d$ its apparent transport constant. The enzyme reaction rate may be expressed as

$$(13) \qquad\qquad V_e = \frac{V_m S^i(m)/K_S}{1 + S^i(m)/K_S + Q^i(m)/K_Q}$$

where V_m, K_S and K_Q are the maximum rate and the Michaelis constants for substrate and product, respectively.

It is convenient to rewrite equations (11), (12) and (13) in dimensionless form by defining scaled substrate and product concentrations, as well as scaled transport constants, namely

$$s_\sigma = \frac{S^i(m)}{K_S} \qquad s_o = \frac{\tilde{S}^o(o)}{K_S}$$

(14)
$$q_\sigma = \frac{Q^i(m)}{K_Q} \qquad q_o = \frac{\tilde{Q}^o(o)}{K_Q}$$

$$h_d^{\bigstar} = \frac{\tilde{h}_d K_S}{V_m} \qquad h_d^{\bigstar'} = \frac{\tilde{h}_d' K_Q}{V_m}$$

If substrate diffusion is "slow" whereas product diffusion is not, there may exist a coupling between substrate diffusion, electric partition of charged ligands and reaction, namely

(15)
$$h_d^{\bigstar}(s_o - \Pi^z s_\sigma) - \frac{s_\sigma}{1 + s_\sigma + q_\sigma} = 0$$

This implies that there exists a steady state dor s_σ and a quasi-equilibrium for q_σ which is nearly identical to q_o (Figure 10A).

Alternately if it is the product which diffuses slowly whereas the substrate diffuses rapidly, there may exist a coupling between product diffusion and reaction (Figure 10B), namely

(16)
$$\frac{s_\sigma}{1 + s_\sigma + q_\sigma} - h_d^{\bigstar'}(\Pi^{\lambda z} q_\sigma - q_o) = 0$$

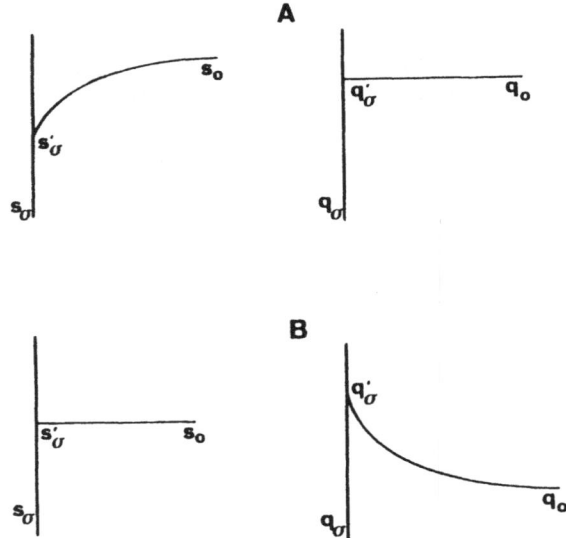

Fig. 10. Coupling between the diffusion of only one ligand and the enzyme reaction.
A / Coupling between substrate diffusion, electric partition and enzyme reaction. The situation in this scheme is the one described by equation (15). The diffusion of the product (right) is assumed to be very fast with respect to that of the substrate in such a way that $q_o = q'_\sigma$.
B / Coupling between product diffusion, electric partition and enzyme reaction. The situation in this scheme is the one described by equation (16). The diffusion of the substrate is very fast and such that $s_o = s'_\sigma$.

In this case it is assumed that it is s_σ which is close to s_o.

If $z = \lambda = 1$ the expression of the partition coefficient at the surface of the membrane may be shown to be (Ricard et al , 1981)

(17)
$$\Pi = \frac{\Delta^-}{S^o(m) + Q^o(m)}$$

where Δ^- is the "concentration" or the density of fixed negative charges. If one wants to study the effect of a variation of substrate or product concentration in the reservoir on the value of the electric partition coefficient one has to maintain constant the degree of advancement of the reaction. This implies that the concentration of substrate and product are changed simultaneously in order to maintain constant the ratio $\rho = Q^o(m)/S^o(m)$.

At constant advancement of the reaction, if there is a coupling between substrate diffusion and reaction the coupling equation (15) may be rewritten as

$$(18) \quad h_d^* s_o \Pi^3 - h_d^* \delta_s \Pi^2 + \{h_d^* s_o(\delta_s + \delta_q) - \delta_s\}\Pi - h_d^* \delta_s(\delta_s + \delta_q) = 0$$

Similarly, the coupling equation (16) at constant advancement of the reaction between product diffusion and reaction may be reexpressed as

$$(19) \quad h_d^* q_o \Pi^3 - h_d^{*'} \delta_q \Pi^2 + \{h_d^* q_o(\delta_s + \delta_q) + \delta_s\}\Pi - h_d^{*'} \delta_q(\delta_s + \delta_q) = 0$$

In either of these equations

$$(20) \quad \delta_s = \frac{\bar{\Delta}}{K_S} \frac{1}{1 + \rho} \qquad\qquad \delta_q = \frac{\bar{\Delta}}{K_Q} \frac{\rho}{1 + \rho}$$

It may be shown that equation (18) may have only one real positive and two imaginary roots. This implies that, for coupling between substrate diffusion and reaction at constant advancement of this reaction, the electric partition coefficient Π monotonically declines as the substrate concentration is varied in the resevoir (Figure 11A). Alternately equation (19) may have three real positive roots. This means that the coupling of product diffusion and reaction, at constant advancement of the reaction, may result in a hysteresis loop of the partition coefficient (Figure 11B). The existence of multiple values of the electric partition coefficient, for a fixed value of product concentration in the reservoir implies that the membrane may store short-term memory. The surface of the membrane may therefore acquire the elementary perception of a change of concentration in the external milieu (Ricard and Noat,1985a)

DYNAMIC BEHAVIOUR OF THE MEMBRANE SYSTEM

We have considered so far the coupling of enzyme reaction with either diffusion of substrate or diffusion of product, but not with both. Let us assume now that there is a coupling between enzyme reaction, diffusion of substrate and diffusion of product.

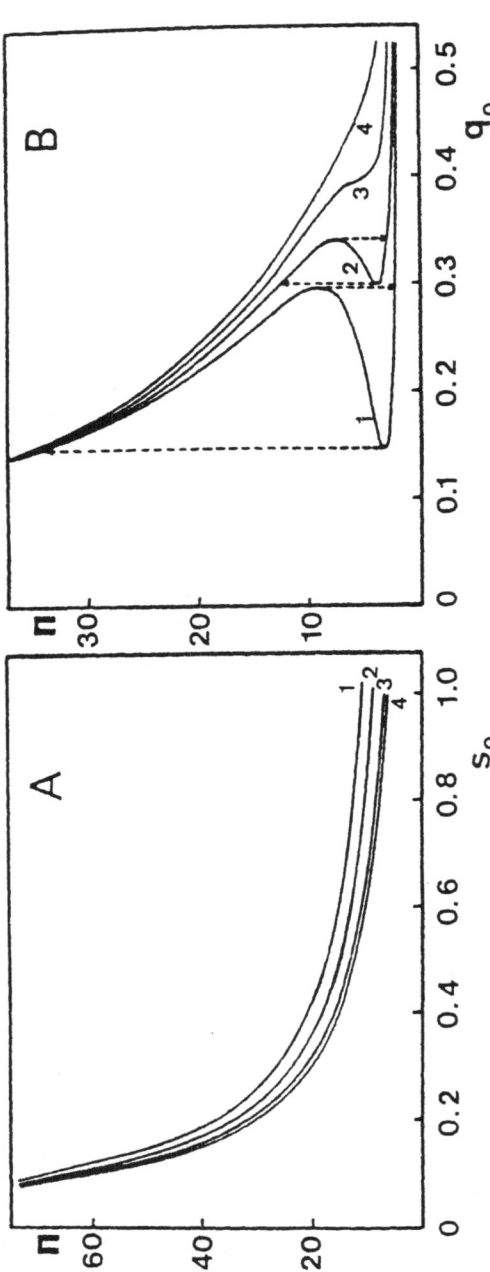

Fig. 11. Variation of the electric partition coefficient, under non-equilibrium conditions and at constant advancement of the reaction, as a function of substrate or product concentration in the reservoir.
A/ Variation of the electric partition coefficient as a function of substrate concentration in the reservoir. A coupling is assumed to occur between substrate diffusion and reaction. Diffusion of the product is assumed to be fast ($q_0 = q'_0$).
B/ Variation of the electric partition of the coefficient as a function of product concentration in the reservoir. A coupling is assumed to occur between product diffusion and reaction. Diffusion of the substrate is assumed to be fast ($s_0 = s_0$).
For panels A and B the transport constant is increased in curves 1 to 4.

Coupling equations (15) and (16) simultaneously hold. Defining scaled influxes and outfluxes of substrate and product as

(21)
$$J_{so}^* = h_d^* s_\sigma \qquad J_{si}^* = h_d^* s_o$$

$$J_{qo} = h_d^{*'} q_\sigma \qquad J_{qi}^* = h_d^{*'} q_o$$

the system of coupling equations may be expressed as

(22)
$$f(s_\sigma, q_\sigma) = 0 = J_{si}^* - \Pi^z J_{so}^* - v_e$$

$$g(s_\sigma, q_\sigma) = 0 = v_e - \Pi^{\lambda z} J_{qo}^* + J_{qi}^*$$

where v_e is the scaled enzyme reaction rate.
The expression of this enzyme reaction rate may be cancelled out from these equations and one finds

(23)
$$\Pi^{\lambda z} J_{qo}^* + \Pi^z J_{so}^* - (J_{si}^* + J_{qi}^*) = 0$$

Clearly there may exist only one value of the partition coefficient for any substrate or product concentration in the reservoir and no hysteresis may occur when the system is fully coupled as described by the system of equations (22).

Another important question is to know whether the membrane system described by equations (22) may exhibit homeostasis or instability when perturbed from its steady state. The system will be defined as homeostatic if it returns back to its initial steady state after a slight perturbation. Alternately the system will be unstable if it diverges from the steady state upon the perturbation. The motion of the system in the close vicinity of the steady state may be analyzed by the phase-plane technique (Minorsky, 1964; Pavlidis, 1973; Nicolis and Prigogine, 1977; Heinrich et al, 1977). The scaled substrate and product concentration, s and q, in the matrix may be expressed from the corresponding steady state concentration s_σ and q_σ plus the deviations x_s and x_q about this steady state, namely

(24)
$$s = s_\sigma + x_s \qquad q = q_\sigma + x_q$$

The dynamics of the system in the vicinity of the steady state may be studied by the analysis of the variational system

$$\frac{d}{dt} \begin{bmatrix} x_s \\ \\ \\ x_q \end{bmatrix} = \begin{bmatrix} \dfrac{\partial f}{\partial s} & \dfrac{\partial f}{\partial q} \\ \\ \dfrac{\partial g}{\partial s} & \dfrac{\partial g}{\partial q} \end{bmatrix} \begin{bmatrix} x_s \\ \\ \\ x_q \end{bmatrix} \qquad (25)$$

obtained by expanding deviations x_s and x_q in Taylor series. The elements of the Jacobian matrix may be found to be of the form (Ricard and Noat, 1985b)

$$\frac{\partial f}{\partial s} = -z \; \Pi^{z-1} J_{so}^{\star} \; \frac{\partial \Pi}{\partial s} - \Pi^{z} \; \frac{\partial J_{so}^{\star}}{\partial s} - \frac{\partial v_e}{\partial s}$$

$$\frac{\partial g}{\partial s} = \frac{\partial v_e}{\partial s} - \lambda z \; \Pi^{\lambda z -1} J_{qo}^{\star} \; \frac{\partial \Pi}{\partial s}$$

(26)

$$\frac{\partial f}{\partial q} = -z \; \Pi^{z-1} J_{so}^{\star} \; \frac{\partial \Pi}{\partial q} - \frac{\partial v_e}{\partial q}$$

$$\frac{\partial g}{\partial q} = \frac{\partial v_e}{\partial q} - \lambda z \; \Pi^{\lambda z-1} J_{qo}^{\star} \; \frac{\partial \Pi}{\partial q} - \Pi^{\lambda z} \; \frac{\partial J_{qo}^{\star}}{\partial q}$$

The dynamic behaviour of the variational system (25) rests on the respective expressions of the trace T_j and the determinant Δ_j of the jacobian matrix as well as on the difference $T_j^2 - 4\Delta_j$. In the absence of any electric repulsion effect ($\Pi = 1$), one may show that $T_j < 0$, whereas $\Delta_j > 0$ and $T_j^2 - 4\Delta_j > 0$. This means that the system may have only a stable node. When perturbed from its initial steady state, the system returns back to the same steady state and therefore exhibits homeostasis (Figure 12).

If alternately electric repulsion effects by the fixed charges

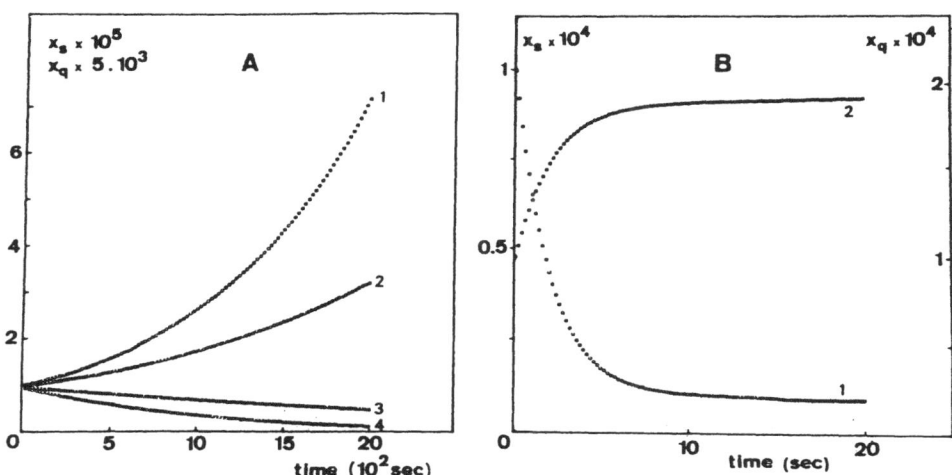

Fig. 12. Stability and instability of substrate and product concentration in a membrane.
A / Stability and instability in the long-term. Curves 1 to 4 show
stabilization of the variational system (24) as a function of time
when ionic strength in the bulk phase is increased. The deviation
x_s is proportional to x_q. At high ionic strength (curve 4) the
system is stable and exhibits homeostasis. For strong electrostatic
repulsion (curve 1) the system is unstable and drifts from its initial steady state.
B / Dynamic behaviour of the system in the short-term. In the short-
term x_s (curve 1) is not proportional to x_q (curve 2). Curves 1
and 2 correspond to the conditions of destabilization by low ionic
strength illustrated by curve 1 of panel A.

of the membrane occur, the trace T_j may be positive, negative or
null, the determinant Δ_j may be either positive or negative and
$T_j^2 - 4\Delta_j$ is always positive. Therefore when electric repulsion
effects occur, the system may exhibit a stable node ($T_j < 0$,
$\Delta_j > 0$) or a saddle point ($T_j > 0$, $\Delta_j < 0$ or $T_j > 0$, $\Delta_j < 0$).
Electric repulsion effects may introduce instability in the system.
Increasing the ionic strength tends to suppress electric repulsion
effects and therefore tends to stabilize the system (Figure 11).

    Instability of the system may be viewed as a device which

allows propagation of a signal along the membrane. A local perturbation may arise in the external milieu, which results in a local change of a reactant concentration in the membrane. If this concentration change is not stabilized rapidly, but is amplified by electric
repulsion effects, it propagates by diffusion. Owing to destabilization, a random localized perturbation may be detected at a remote
distance of the region where it has taken place.

CONCLUSIONS

Several general ideas and conclusions may be derived from the
above theoretical and experimental results.

Under quasi-equilibrium conditions electrostatic repulsion of
substrate by the fixed negative charges of the polyelectrolyte
matrix generates apparent positive co-operativity of the enzyme
bound to this matrix. Alternately electrostatic attraction of the
substrate results in the apparent negative co-operativity . If the
enzyme is heterogeneously distributed in the matrix, the bound
enzyme displays an apparent negative co-operativity or a mixed
positive and negative co-operativity. All these effects must be
highly controlled by the ionic strength of the bulk phase.

Acid phosphatase bound to plant cell wall displays a kinetic
behaviour which is quite compatible with the above theoretical predictions. Acid phosphatase displays in free solution a classical
Michaelis-Menten behaviour, but an apparent negative co-operativity
when bound at the surface of plant cell walls. This begative co-
operativity is suppressed when raising the ionic strength.

When diffusional resistances occur at the surface of a charged
membrane, the concept of electric partition coefficient may still
be applied under non-equilibrium conditions. If a coupling exists,
at constant advancement of the reaction, between product diffusion
and reaction, multiple values of the electric partition coefficient
may exist at the surface of the membrane. This implies that this
membrane may store short-term memory and may react differently depending on the product concentration is increased or decreased in
the reservoir. Therefore, owing to the coupling between product diffusion, reaction and electric partition, the surface of the cell may
have an elementary perception of the external melieu (Hervagault et
al, 1984).

A random perturbation of the steady state of the whole system
may create a destabilization of that system which may diverge from
its initial steady state. This destabilization appears as a direct
consequence of electric repulsion of substrate and product. The
amplification of a concentration change may then propagate along the
surface of the membrane. This conduction of a signal may be viewed
as a device which allows sensing of the external milieu.

Last but not least, the complex dynamic properties of the system are not due to a complex macromolecule but rather to the interplay of ligand diffusion, enzyme reaction and electric ligand repulsion effects.

REFERENCES

Crasnier, M., Noat, G. and Ricard, J.(1980a) Plant Cell Environ. $\underline{3}$, 217-224.

Dixon, M. and Webb, E.C. (1979) Enzymes, Longman, London.

Engasser, J.M. and Horvath, C. (1975) Biochem. J. $\underline{145}$, 431-435.

Heinrich, R., Rappoport, S.M. and Rappoport, T.A. (1977) Prog. Biophys. Mol. Biol. $\underline{32}$, 1-82.

Hervagault, J.F., Breton, J., Kernevez, J.P., Rajani, J. and Thomas , D. (1985) This volume.

Hill, T.L. (1977) Free Energy Transduction in Biology. Academic Press, New York.

Hill, T.L. and Chen, Y. (1970) Proc. Natl. Acad. Sci. USA $\underline{66}$, 607-614.

Hill, T.L. and Chen, Y. (1971) Biophys. J. $\underline{11}$, 685-710.

Minorsky, N. (1969) Theory of Non-Linear Control System, Mac-Graw Hill, New York.

Nicolis, G. and Prigogine, I. (1977) Self-Organization in Non-Equilibrium Systems. John Wiley and sons, New York.

Noat, G., Crasnier, M. and Ricard, J.(1980b) Plant Cell Environ. $\underline{3}$, 225-229.

Pavlidis, T. (1973) Biological Oscillators : Their Mathematical Analysis, Academic Press, New York.

Ricard, J. , Noat, G. Crasnier, M. and Job, D. (1981) Biochem. J. $\underline{195}$, 357-367.

Ricard, J. and Noat, G. (1985a) J. Theor. Biol., in press.

Ricard, J. and Noat, G. (1985b) J. Theor. Biol., in press.

PHOTOBIOCHEMICAL MEMORY

J.F.Hervagault[1], J. Breton[1], J.P. Kernevez[2], J. Rajani[2]
and D. Thomas[1]

1. Laboratoire de Technologie Enzymatique,ERA n°338 CNRS
2. Département de Mathématiques Appliquées
Université de Technologie de Compiègne, 60206 Compiègne
France

INTRODUCTION

The behavior of biocatalysts has been studied for a long time solely from the point of view of their time-dependence. Recently a more relevant and more realistic approach has consisted in coupling pure catalytic functions, with various phenomena intrinsically linked to space, viz. diffusion and/or partition of metabolites, and convection (flow rate, electric field, ...).

As far as autocatalytic steps are involved in the process, properties qualitatively different from those observed in solution, e.g. multistability, time and space structurations, are likely to occur, as predicted theoretically by the Brussels group (Glansdorff and Prigogine, 1971). Concerning multistability, numerous experimental evidences for either kinetic or electrochemical memory in diffusion-reaction coupled enzyme systems, have been reported (Naparstek et al, 1974; Thomas et al, 1977).

Nevertheless, in the framework of this spatiotemporal conception, still little attention has been paid to radiation phenomena which are also essentially space-dependent. Among them, the first phenomenom is obviously the light radiation whose role is fundamental in numerous biological processes, such as photosynthesis. Influence of light radiation (laser beam) on a non linear chemical reaction has been already published (Creel and Ross, 1976).As here, light absorption acts only as the driving force, the approach is quite different from the one presented in that Paper.

157

Thylakoid chloroplasts are immobilized on the surface of an artificial membrane lying on the bottom of a continuously feeded reactor (C.S.T.R.) . The membrane is illuminated perpendicularly with a red light. Thus, the radiation first traverses the substrate solutions (DCPIP) before exciting the thylakoids. Under suitable conditions, the outside substrate concentration as to the solution thickness within the reactor and the flow rate being held fixed, the steady state DCPIP concentration is measured for a light intensity within a defined range of values (I_c). Depending on whether at time zero, intensity is lower or upper than I_c, the steady state concentration will be either close to zero or close to the outside concentration, showing thus the existence of multiple steady states.

Such a multistability is classically explained in term of coupling between a transport process (flow rate) and a non-linear kinetic one. The original feature of the present system lies in the fact that the autocatalytic step emanates from the association of two phenomena intrinsically "linear" , viz. the decrease in intensity (when light traverses the absorbing solution), governed by an exponential law (Beer-Lambert), and the thylakoid activity as a function of substrate concentration, following hyperbolic relation.

In order to explain qualitatively the observed experimental results, a simple model is described and analyzed , model in which multistability may occur. A rapid study of two other models, close to the former one, lead to the conclusion that the spacial partition between light absorbance and a photo-assited reaction improve the efficiency (in terms of memorizing ability) of such interacting processes.

MATERIAL AND METHODS

Chloroplast thylakoids were prepared from fresh lettuce (Latuca sativa, var. romaine) according to the method of Epel and Neumann (1973), modified by Thomasset et al (1982).

Thylakoid membranes were produced by using a previously described method (Cocquempot et al, 1979), with minor alterations.

All experiments are performed in a sorbitol buffer 330 mM, pH 7.6, plus HEPES 50 mM, EDTA 0.4 mm, Na_2HPO_4 0.15 mM, $MnCl_2$ 0.1 mM, $MgCl_2$ 5 mM and $CaCl_2$ 50 mM.

The experimental device consists in a CSTR, thermostated at 20°C. The thylakoid membrane lies on the bottom of the reactor, filled with DCPIP (electron acceptor), the "substrate" of the reaction. Both the reactor volume and the inlet flow-rate are held

constant. The system is illuminated perpendicularly with a red
light (transmittance beyond 580 nm). Light intensities are measured
with a Kipp and Zonen pyranometer (CM 10 solarimeter). Under this
lightning condition, the activity-expressed in terms of DCPIP disappearance (reduction by thylakoids)- is about 80 per cent of the
activity measured in the presence of white light. As DCPIP exhibits
an absorption peak centered around 600 nm (ε_{600} = 17.9 mM^{-1}.cm^{-1}),
part of the incident light intensity is catched when traversing
the solution, before exciting thylakoids at the solution—membrane
interface (see scheme, model I).

   Outlet DCPIP concentration is followed spetrophotometrically
at 670 nm (ε_{670} = 9.9 mM^{-1}.cm^{-1}).

EXPERIMENTAL RESULTS

   When carried out under batch conditions, the measurements of
activity as a function of DCPIP concentration, for two incident
light intensities, give the results shown on Figure 1. As the concentration is increased, an apparent activity inhibition effect due
to an excess of DCPIP, is observed.

   Considering now the same system working under open conditions,
with fixed outside DCPIP concentration (0.16 mM) and flow rate
(7 cm^3.h^{-1}), the actual steady state will be reached when transport
will balance the reaction. Graphically (Thomas et al, 1977), it is
defined as the intercept of the previously obtained curves (reaction term) with a straight line taking into account flow rate
and substrate outside concentration (transport). Under a light intensity equal to 800 W/m^2, a single steady state exists. On the
other hand, when light intensity is decreased down to 300 W/m^2 (I_c),
three steady states can be expected, one of them being unstable.

   Thus, the qualitative behavior of this sytem suggests that
the DCPIP steady state concentration, as to activity, will depend
on whether at time zero —start of the illumination at I_c- the
system would have been illuminated with an intensity upper, e.g.
800 W/m^2, or lower, e.g. nil, than I_c. Results of experiments
carried out in that way are given in Figure 2. After illumination
at 800 W/m^2, the DCPIP steady state concentration, equals 0.018 nM,
regardless its initial concentration within the reactor. After a
sudden decrease in intensity down to 300 W/m^2, the new steady
state is reached, corresponding to a concentration of 0.021 mM.
On the other hand, when starting the experiment in the absence of
light, the steady state concentration equals the outside reservoir
concentration , viz. 0.16 mM. Increasing intensity up to 300 W/m^2,
will drive the system to a new steady state for which the DCPIP
concentration equals 0.133 mM.

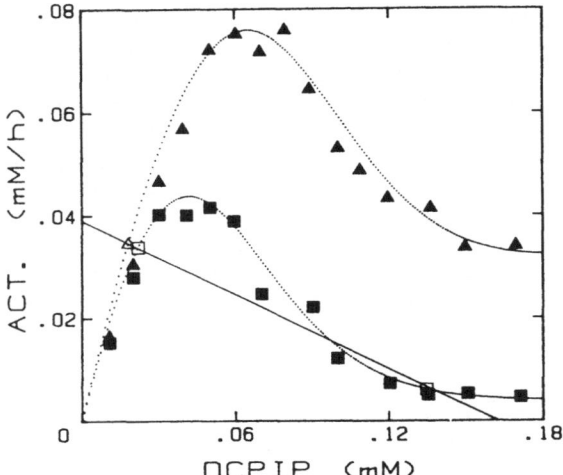

Fig.1. Thylakoid membranes activity under batch conditions as a function of the DCPIP concentration. Activity is expressed in term of
DCPIP consumption as a function of time (mM.h^{-1}). Experiments are
carried out in a 30 cm^3 substrate volume (solution thickness equal
to 2.5 cm). Incident light intensities equal 800 W/m^2 (▲) and
300 W/m^2 (▫). The straight line has been drawn in accordance with
parameter values as follow : 0.16 mM for the outside DCPIP concentration and 7 cm^3 h^{-1} for flow rate. △ (800 W/m^2) and □ (300 W/m^2):
stable steady states of the system when working under open conditions.

     The different steady state concentrations -linked to different
activities- depending on the history of the system, are quantitatively in excellent agreement with the former predictions. Let us note
that during the evolution experiments, no significant loss of thylakoid activity has been tested for . In the same way, the occurrence of multiple steady states, when outside DCPIP concentration is
increased and decreased, under a constant incident light intensity,
has also been demonstrated (Breton et al, submitted for publication).

THEORETICAL INTERPRETATION AND DISCUSSION

     Let us consider the transformation S $\xrightarrow{h\nu}$ P , e.g. the photo-
reduction of an electron acceptor by thylakoids. Under a suitable

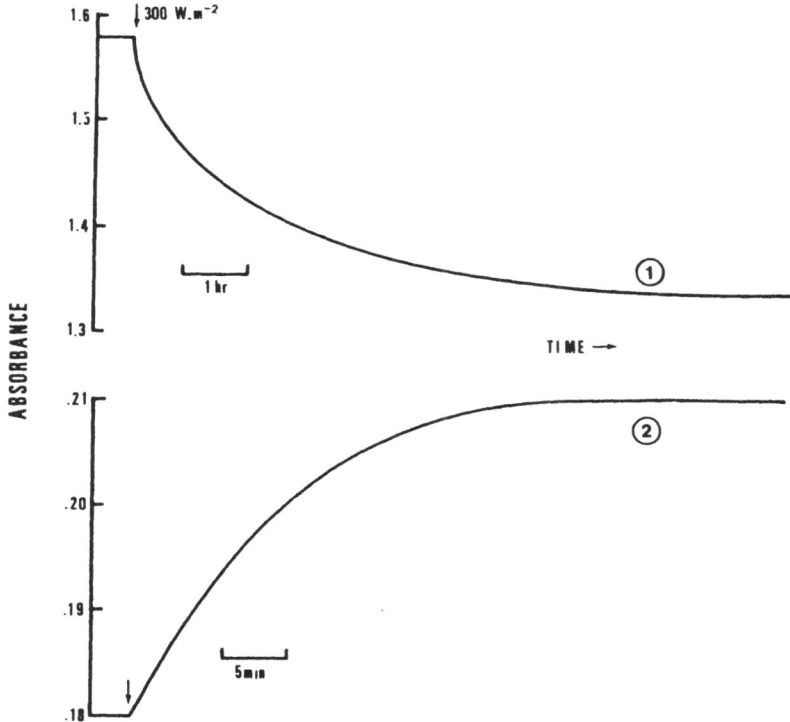

Fig.2. Evolution of the outlet DCPIP concentration under open conditions . Absorbance is measured at 670 nm (ε_{670} = 9.9 mM^{-1}). Light intensities before transition at 300 W/m^2 ($\downarrow$) were 800 W/m^2 and zero (curve 2) , respectively . For details, see text.

excitation light wavelength, the rate of reduction of S will be proportional to the incident light 1. For given wavelength and intensity, the net rate of transformation as a function of S-experimentally observed- is of the same shape as either the standard reaction rate for enzymic reactions involving a single substrate, or the Langmuir adsorption isotherms (rectangular hyperbola). Although the mathematical expressions are similar, no proper interpretation of the underlying mechanism is intended.

Increasing incident light intensity I, will lead to a series of affine hyperbolic curves of activity. Thus, the net rate of consumption of S is defined as :

(1) $$\frac{dS}{dt} = -I\ f(S) \quad \text{with} \quad f(S) = V.S/K + S$$

where V is the maximal rate consumption and K is the concentration
of S for which $f(S) = V/2$.

Such a description assumes the process to take place in an
ideal homogeneous and isotropic medium. Obviously, these conditions
are unlikely to occur under in vivo situations. Therefore, a more
realistic description would take into account at least two major
components, namely :
- Space organization and/or compartmentalization which are
  fundamental features of the cell structure.
- Open working conditions allowing exchanges of matter, energy
  and information with the outside environment.

The experimental situation studied in that paper, can be described in a realistic way by model I (see scheme).

Here the consumption of S as a function of time, will be the
algebraic sum of its supply from flow-rate and its consumption
by the light-dependent reaction

$$(2) \qquad \left(\frac{\partial S}{\partial t}\right) = \left(\frac{\partial S}{\partial t}\right)_F + \left(\frac{\partial S}{\partial t}\right)_{reaction}$$

If substrate molecules do not absorb at the incident light wavelength, the global behaviour of the system will be qualitatively
similar to the simple one described by Eq.(1). Consider now the
more attractive alternative case where substrate is absorbed at
that wavelength. Indeed, as the radiation traverses the reservoir,
its intensity decreases. This change in intensity is governed
by the well-known Beer-Lambert law :

$$(3) \qquad -dI(x) = I(x)\ \varepsilon S\ dx$$

where $I(x)$ is the light intensity at a certain depth x, and ε is
the extinction coefficient for substrate S. Upon integration, equation (3) becomes

$$(4) \qquad I(x) = I_o \exp(-\varepsilon x\ S) \quad \text{where } I_o = I(o)$$

In this model, the true intensity I occurs at $x = L$ and

$$(5) \qquad I(L) = I = I_o \exp(-k'\ S)$$

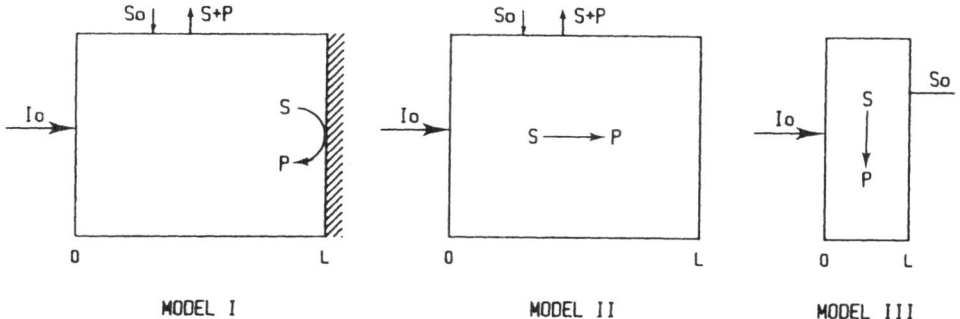

MODEL I   MODEL II   MODEL III

Scheme. Description of models

Model I - An active film on the surface of which the photoassisted reaction takes place, lies on the bottom of a C.S.T.R. and incident perpendicular light I_O traverses the solution of substrate. L is the depth of the solution, S_O and S are the in and outlet concentrations, respectively. The reactor with a volume v is fed with substrate at a flux rate D. This model is related to the experimental situation.

Model II - The catalytic reaction takes place throughout the solution, with all other features as in model I.

Model III - An inert artificial membrane,whith a thickness L, containing a uniform distribution of catalytic sites immobilized therein is coated along a glass plate and exposed to a batch containing substrate at a constant concentration S_O.

where $k' = \varepsilon L$ and $I_O = I(o)$. Therefore, according to equ.(2) , the system is governed by

$$(6) \qquad \frac{ds}{dt} = s_o - s - \lambda \exp(-ks) \frac{s}{1 + s}$$

with $\lambda = I_o \sigma$ where $\sigma = V.v/(K.D)$ and $k = \varepsilon LK$. s_o and s are normalized concentrations : $s_o = S_O/K$ and $s = S/K$. Characteristic time for diffusion (v/D) is taken as unit time. The reaction term R(s) comprises the product of the term(s/(1+ s)) by the decreasing exponential function : exp (-ks).

Figure 3 shows the resulting function R(s) which is similar in
shape to functions observed with substrate inhibited kinetics, the
"inhibitory effect" depending mainly on the value of the parameter
k (= εLK).

Let us note that k = 0 (no inhibition) corresponds to situations
where either substrate does not absorb (ε =o) or space is not taken
into consideration (L = o).

The classical rate of a photoassisted catalytic reaction —as
described by equ. (1)— is greatly modified when absorbance of the
incident light by the substrate throughtout the reactor is taken
into account. Such a non linear reaction term does not follow the

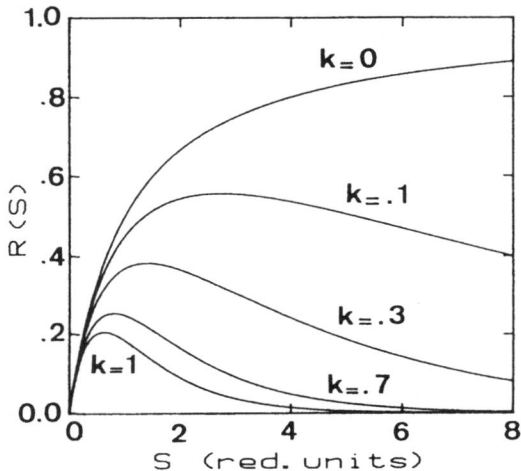

Fig.3. Resulting R(s) function $\left[\exp(-ks) \dfrac{s}{1+s} \right]$ for various
values of the k parameter. The abcissa s is expressed in normalized
units (s=S/K). The higher the k value, the more important the inhibitory effect. R(s) is hyperbolic for k = o (either ε or L equal
to zero).

Lechatelier–Braun law and exhibits an autocatalytic branch in its
right–handed part. Previous works have abundantly demonstrated that
the coupling of such non–linear reaction terms are able to give rise
to multiple steady states when coupled with transport processes
(Aris and Keller, 1972). Indeed, it is trivial to check that the
zero–dimensional system (6) possesses multiple steady state solutions of

$$(7) \qquad \frac{1}{\lambda} (s_o - s) = \exp (-ks) \frac{s}{1+s}$$

when either s_o or λ vary.

In Figure 4a the steady state solutions s are represented for
equation (7) when λ is taken as the control parameter. The existence of such an S– shaped curve of solution leads to the appearance
of an hysteretic behaviour when λ is increased and decreased.

Figure 4b represents the projection of the catastrophe surface
in the parametric plane (s_o, λ). The direct involvment of this
multistability relies on the fact that such a simple system has a
memorizing ability. Considering a couple (s_o, λ) located within the
catastrophe surface, and a given initial condition for s, e.g.
s(o) = 0, the (stable) steady state will depend on whether λ (or
so) will be reached by an increase, e.g. $\lambda(o) = o$, or decrease
(in λ), e.g. $\lambda(o) \gg \lambda(\text{beyond } T_2)$. In the first case, the steady
state will lie on the upper branch of solutions (reaction control-
led) and in the second one, it will lie on the lower branch (diffusion controlled).

In other terms, the activity will be quantitatively different
in accordance with time zero when the system was placed in the
dark or illuminated. In a certain extent we are faced to an all or
nothing situation.

Other related models and discussion

The same study dealing with the effect of light radiations on
a phosphoassisted reaction could be carried out with experimental
set up designed in a different manner. Although these approaches
are still under progress, a rapid mathematical analysis is proved
useful (see scheme, models II and III).

In model II , where the reaction takes place in the whole
volume of reactor, the global reaction rate is defined for any
thickness x, between o and L.

(8)
$$I = I_o \int_o^L \exp(-ks)dx = I_o \left[1 - \exp(-ks)\right]/ks$$

leading to,with the same transformations as in equ. (6)

(9)
$$\frac{ds}{dt} = s_o - s - \lambda \left\{\left[1 - \exp(-ks)\right]/ks\right\} \frac{s}{1+s}$$

Finally, in the membrane model III, consumption of S is directly linked to its diffusion within the structure. Consequently S will be a function of both time t and its position x in the membrane.

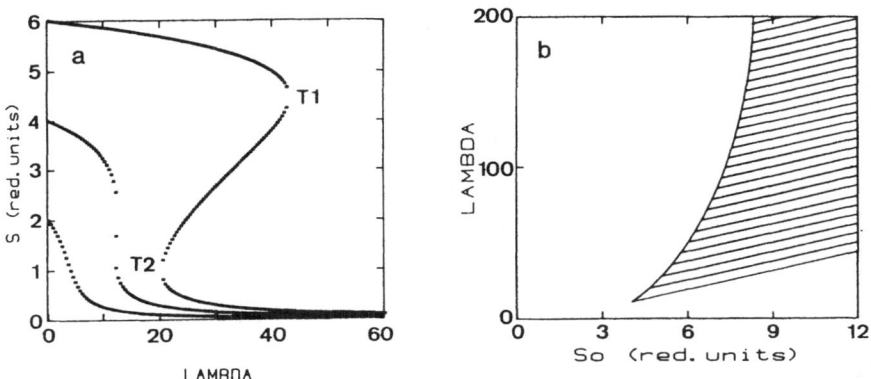

Fig.4. a/ Representation of the steady state s (from equ.(7)) as λ varies, for various s_o values. k = 0.7. For large enough values of s_0 ($s_0 > 4$), curves show the existence of multiple steady states regions. They are bound by the two turning points T_1 and T_2 for which the observed steady states are discontinuous functions of λ. b/ Representation of the projection of the catatrophe surface (dashed area) in the parametric plane(s_0, λ), for k=0.7.

(10) $s = s(x,t)$ $o \leqslant x \leqslant L$ $t \geqslant o$

The transport term being given by Fick's second law, model III is ruled by the following partial differential equation(Kernevez, 1980).

$$(11) \quad \begin{cases} \dfrac{\partial s}{\partial t} = \dfrac{\partial^2 s}{\partial x^2} - \lambda \exp\left[-k \int_o^x s(u)\ du \right] \dfrac{s}{1+s} \\[2em] s(L,t) = s_o \ , \quad \dfrac{\partial s}{\partial x}(o,t) = o \end{cases}$$

where $\lambda = I_o\ VL^2/KD_S$ and D_S is the diffusion coefficient for S through the membrane. The exponential term comes from the solution of

$$(12) \quad \frac{\partial I}{\partial x} + kIs = o \qquad I(o,t) = I_o$$

which is

$$(13) \quad I(x,t) = I_o \exp\left[-k \int_o^x s\ (u,t)\ du\right]$$

These two models exhibit a qualitatively identical behavior, compared to model I.

As an illustration, S-shaped curves of solutions for various depths calculated from model III, with varying λ, are shown in figure 5.

A similar modelling but in which only electron transport is related to the incident light intensity (no occurrence of non linearity) has been proposed recently (Howell and Vieth, 1982).

The feasibility of acquiring multiple steady states with any of the three models being granted, a quantitative comparison is of interest : considering the likelihood for multistability to occur, as an efficiency criterion, thus the larger the multistability region and the lower the value of the control parameters, the more likely the appearence of the phenomenon. This approach led us to observe that the above defined efficiency decreases from model I to model III (I → II → III). It is clear from this result that efficiency is directly linked to the proper spatial arrangement of any model: substrate distribution is homogeneous in I and II, whereas it is heterogenous in III. Concerning activity, it is homogeneously distributed solely in model I. Thus, it follows that spatial partition and/or compartmentalization between light absorbance and reaction improve the efficiency of such coupled systems.

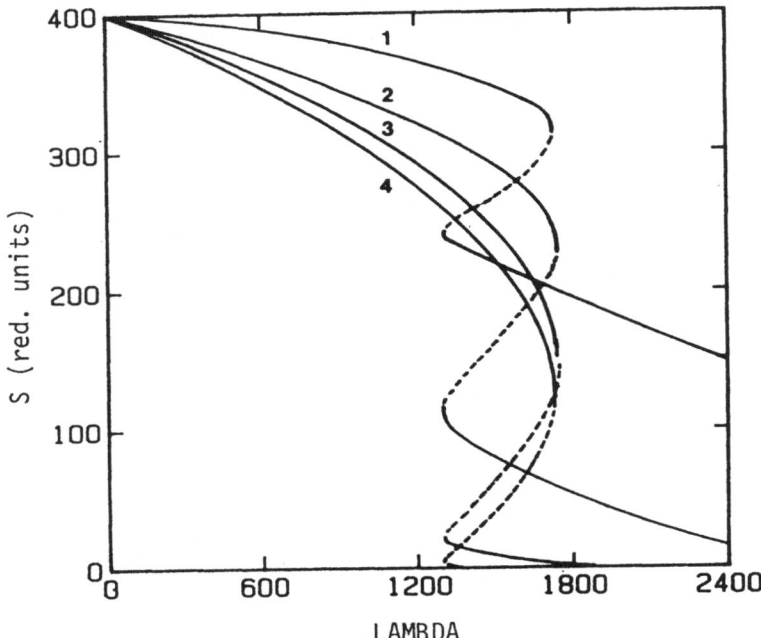

Fig.5. Calculated branches of steady state solutions of equation (10) as λ varies, and for k = 0.04.

$$\begin{cases} -\dfrac{\partial^2 s}{\partial x^2} + \lambda \exp\left[-k_0 \int_0^x s(x)\,dx\right] \dfrac{s(x)}{1 + s(x)} = o \\[3mm] \dfrac{\partial s}{\partial t}(o) = o \qquad s(L) = s_0 \end{cases}$$

This result has been obtained by using the Kubicek method of continuation (Kubicek, 1976). Curves labelled 1, 2, 3 and 4 represent the concentrations s(4L/5), s(3L/5), s(2L/5) and s(L/5) respectively.
(——) stable and (---) unstable steady states.

    A plausible transposition of the above proposed models to in vivo situations, in the particular case of photosynthetic systems, may consist of a spatial arrangement of photoreceptors functionally linked to the protein pigment entities, the receptors acting as filters between the incident light and centers carrying out the electronic transfer.

·   This new and certainly promising way of looking at it is not restricted to photosynthesis but may be relevant for any light dependent processes such as metabolic inductions by phytochrome or retinal physiology.

Finally, considering any phenomenon linked to radiations, e.g. neutronics, its space dependent connection with catalytic processes should lead to the discovery of unexpected properties.

REFERENCES

Aris, R. and Keller, K.M. (1972) Proc. Natl. Acad. Sci. USA 69 , 777-779.

Breton, J., Thomas, D. and Hervagault, J.F. (1984) submitted for publication.

Cocquempot, M.F., Thomas, D., Champigny, M.L. and Moyse, A. (1979) Eur. J. Applied Biotechnol. 8, 37-41.

Creel, C.L. and Ross, J. (1976) J. Chem. Phys. 65 , 3779- 3789.

Epel, B.L. and Neumann, J. (1973) Biochim. Biophys. Acta 325, 520-529.

Glansdorff, P. and Prigogine, I. (1971) "Thermodynamic Theory of Structure, Stability and Fluctuations", Interscience, Wiley, New-York.

Howell, J.M. and Vieth, W.R. (1982) J. Mol. Catalysis 16, 245- 298.

Kernevez, J.P. (1980) "Enzyme Mathematics" Ed. Lions, J.L., Papanicolaou, G. and Rockfellar, R.T. ,North-Holland , Amsterdam.

Kubicek, M. (1976) A.C.M. Transactions on Mathematical Software 2 , 98- 107.

Nasparstek, A., Romette, J.L., Kernevez, J.P. and Thomas, D. (1974) Nature 249, 490-491.

Thomas, D., Barbotin, J.N., David, A., Hervagault, J.F. and Romette, J.L. (1977) Proc. Natl. Acad. Sci. USA 74, 5314-5317.

Thomasset, B., Thomasset, T., Vejux, A., Jeanfils, J., Barbotin, J.N. and Thomas, D. (1982) Plant Physiol. 70, 714-722.

SECTION V – DYNAMICS OF METABOLIC PATHWAYS : SELF-ORGANIZATION AND CHAOTIC BEHAVIOUR

FROM EXCITABILITY AND OSCILLATIONS TO BIRHYTHMICITY AND CHAOS IN BIOCHEMICAL SYSTEMS.

A. Goldbeter, J-L. Martiel and O. Decroly

Faculté des Sciences, Université Libre de Bruxelles
CP 231, 1050 Brussels, Belgium

> "Avez-vous visité les hauts laboratoires,
> Où l'on poursuit, de calcul en calcul,
> De chaînon en chaînon, de recul en recul,
> A travers l'infini, la vie oscillatoire ? "
>
> Emile Verhaeren
> Les Forces Tumultueuses, 1902

INTRODUCTION

Spontaneous oscillations are one of the most conspicuous properties of living systems (Winfree, 1980). They are seen in the firing patterns of neurons, in beating heart cells, in the wavelike aggregation of cellular slime molds, in circadian rhythms which can persist under constant external conditions, in hormonal cycles which govern reproduction in mammals, and in the periodic evolution of interacting populations of predators and preys. This list, which is by no means exhaustive, illustrates how rhythmic phenomena occur at all levels of biological organization, with periods ranging from seconds to years.

Oscillations are not restricted to living organisms. They are also observed in purely chemical systems such as the Belousov-Zhabotinsky (Field et al, 1972) or Briggs-Rauscher (Pacault et al, 1976) reactions. In spite of their diversity, chemical and biological oscillations possess common roots and can be comprehended as a process of temporal self-organization. From a thermodynamic point of view, such sustained oscillations represent temporal dissipative structures as they occur beyond a bifucation point, i.e. a critical point of instability of a nonequilibrium stationary state (Prigogine, 1967; Nicolis and Prigogine, 1977). The prerequisites for a sustained oscillations are threefold : they occur in open systems governed by appropriate nonlinear kinetic equations, beyond a critical distance from equilibrium (Prigogine, 1967; Nicolis and Prigogine, 1977).

The sources for nonlinearity are manifold in biology. Kinetic equations become nonlinear as a result of cooperative interactions and feedback processes, be it for ion transport across excitable membranes, control of genetic or enzyme activity, hormonal regulation, or interactions between animal populations. This explains why periodic phenomena are so common in biological systems. The precise mechanisms responsible for the onset of oscillations remain, however, largely unclear for many biological rhythms. Positive feedback plays an essential role in the origin of most rhythmic phenomena. As will be shown below, this view is supported by the analysis of biochemical oscillations which are the best understood at the molecular level (Hess and Boiteux, 1971; Goldbeter and Caplan, 1976; Berridge and Rapp, 1979).

This paper consists of two main parts. In the first, we consider the molecular basis of some well-known biochemical oscillations. These are the periodic operation of glycolysis in yeast and muscle, and the periodic synthesis of cyclic AMP signals which control aggregation of Dictyostelium discoideum amoebae after starvation. The latter system will be considered in particular detail. We shall successively address the mechanism of oscillations and relay of cAMP signals, and will show that the latter phenomenon, which consists in the pulsatory amplification of suprathreshold cAMP pulses, reflects the excitability of D. discoideum cells. Recent experiments have shown that the cells respond and adapt to constant cAMP stimuli, probably through some receptor modification. To account for these various phenomena, an extension of a model proposed by Goldbeter and Segel (1977) for the signalling system will be presented, based on desensitization of the cAMP receptor.

The cAMP signalling system provides an excellent example of the physiological significance of excitability and oscillations at the cellular level. Moreover, it offers a clue as to the role of bifurcations in the control of developmental transitions. We shall show how the model can be used to explain the sequence of transitions no relay-relay-oscillations observed after starvation.

The above examples illustrate how a single positive feedback loop can produce excitable and oscillatory behavior in biochemical systems. The question arises as to what is the effect of a coupling between two instability-generating mechanisms in the same system? In the second part of the paper, we shall analyze a model for a sequence of two autocatalytic enzyme reactions (Decroly and

Goldbeter, 1982) in which the variety of behavioral modes is greatly enhanced. In addition to excitability and simple periodic oscillations, new phenomena appear such as the coexistence between two
stable periodic regimes. We have referred to the latter phenomenon
as birhythmicity (Decroly and Goldbeter, 1982), in analogy with
bistability in which two stable steady states coexist under the
same conditions. The same system can even exhibit a phenomenon of
trirhythmicity in which three stable periodic regimes coexist for
the same set of parameter values.

The interplay between two positive feedback processes can also
produce complex periodic oscillations which resemble the bursting
behavior of nerve cells, and aperiodic oscillations. The latter
phenomenon, referred to as chaos, has received much attention in
physics and chemistry (Vidal and Pacault, 1981) as it introduces an
element of randomness in the behavior of systems governed by deterministic evolution equations.

The prediction that complex oscillations and birhythmicity
originate from the coupling between two autocatalytic reactions
has recently been corroborated by experiments in a chemical system
(Alamgir and Epstein, 1983). We were somewhat surprised to find
complex periodic oscillations, birhythmicity and chaos in the extended model based on receptor modification for the cAMP signalling
system of D. discoideum . Here, the receptor dynamics introduces a
second path for cAMP oscillations. Complex phenomena originate from
the interplay between two parallel oscillatory paths sharing a
single positive feedback process. In conclusion, we shall discuss
the relevance of these results to D. discoideum behavior; in particular , we shall review the evidence that suggests the occurrence
of aperiodic oscillations of cAMP in the mutant FR17 (Durston,
1974), before comparing the likelihood of chaos versus periodic
behavior in biochemical systems.

PERIODIC BEHAVIOR AND EXCITABILITY

Glycolytic oscillations

Since their discovery some twenty years ago (Ghosh and Chance,
1964), glycolytic oscillations have become the prototype of periodic
behavior in biochemistry. The numerous experimental and theoretical
studies devoted to this phenomenon have been reviewed (Hess and
Boiteux, 1971; Goldbeter and Caplan, 1976; Berridge and Rapp, 1979;
Hess and Boiteux, 1968; Pye, 1969). We shall only recall here the
salient features of these oscillations and the basic elements of
their analysis.

Glycolytic oscillations have been observed in intact yeast
cells (Pye, 1969; von Klitzing and Betz, 1970) and in extracts of

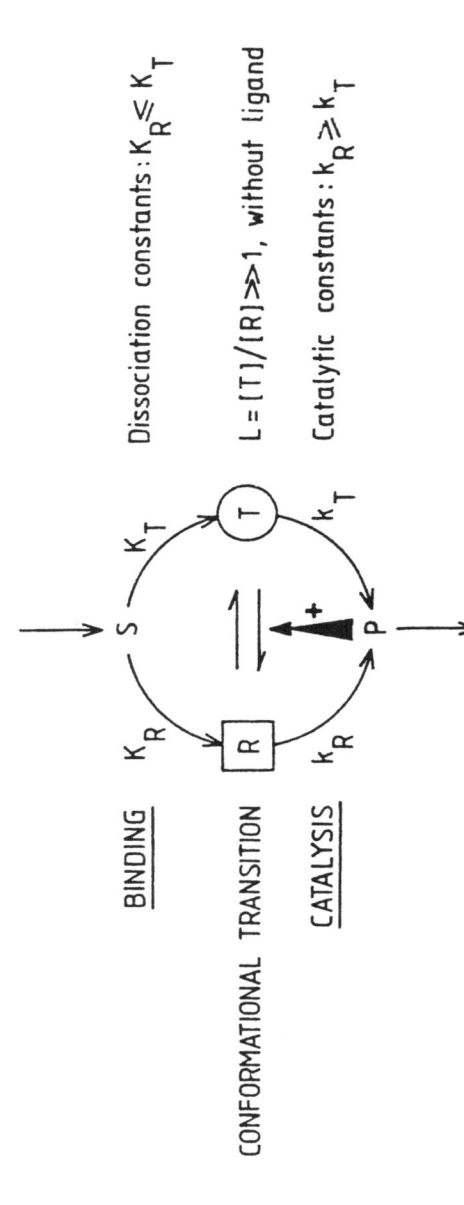

BINDING

CONFORMATIONAL TRANSITION

CATALYSIS

Dissociation constants : $K_R \leqslant K_T$

$L = [T]/[R] \gg 1$, without ligand

Catalytic constants : $k_R \geqslant k_T$

Fig.1. Model of an allosteric enzyme activated by a reaction product. The enzyme consists of several subunits which undergo a concerted transition (Monod et al, 1965) between the T and R states. The model applies to phosphofructokinase which produces glycolytic oscillations ; the product ADP is indeed a positive effector of the enzyme.

yeast (Hess and Boiteux, 1971, 1968; Pye, 1969) and muscle (Frenkel, 1968; Tornheim and Lowenstein, 1974). Depending on the substrate input, the period of the oscillations ranges from 8 to 3 min in yeast extracts (Hess et al, 1969); it is shorter in yeast cells (Pye, 1969; von Klitzing and Betz, 1970) and longer in muscle extracts (Tornheim and Lowenstein, 1974) by up to one order of magnitude. The oscillations result from the positive feedback exerted by a reaction product (ADP, or the related product AMP in yeast, and fructose 1,6-bisphosphate in muscle). All glycolytic intermediates oscillate with the same frequency but with different phase relationships(Hess and Boiteux, 1971). The role of the various intermediates in the mechanism of oscillations can be demonstrated in phase-shift experiments : only the controlling metabolites such as ADP in yeast (Hess and Boiteux, 1968; Pye, 1969) and fructose 1,6-bisphosphate in muscle (Tornheim and Lowenstein, 1974) produce such phase-shifts. It is intriguing, in this respect, that the addition of fructose 2,6-bisphosphate does not cause significant phase-shifts of glycolytic oscillationsin yeast extracts (Hess, personal communication). This metabolite has recently been identified as a most potent activator of phosphofructokinase (Hers and van Schaftingen, 1982), also active on the yeast enzyme. The lack of major effect of this metabolite is even more intriguing, given that the addition of positive effectors of phosphofructokinase such as ammonium ions in yeast instantaneously suppress the oscillations (Hess and Boiteux, 1968; Hess et al, 1969). A similar phenomenon occurs upon addition of negative effectors such as citrate in muscle Frenkel, 1969). These observations suggest that the cooperative properties of phosphofructokinase are an essential element in the mechanism of oscillations (Hess et al, 1969), since both positive and negative effectors can transform the kinetics of allosteric enzymes from sigmoidal to Michaelian.

Early models for glycolytic oscillations (Higgins, 1964; Sel'kov, 1968) were already based on the positive feedback exerted by a reaction product of phosphofructokinase. Later on, an allosteric model for a product-activated enzyme was developed (Goldbeter and Lefever, 1972; Goldbeter and Nicolis, 1976) in the frame of the concerted transition theory of Monod, Wyman and Changeux (1965). This model yields qualitative and quantitative agreement with most experiments on glycolytic oscillations in yeast extracts, both for a constant substrate injection rate – such as used in experiments with yeast cells and extracts – and for a periodic substrate input (Boiteux et al, 1975).

For a dimeric enzyme, the time evolution of the substrate (α) and product (γ) normalized concentrations is governed in this model by the kinetic equations (Goldbeter and Lefever, 1972; Goldbeter and Nicolis, 1976; Venieratos and Goldbeter, 1979) :

(1) $d\alpha/dt = v - \sigma\phi$, $d\gamma/dt = q\sigma\phi - k_s\gamma$

with $\phi = \alpha(1 +\alpha)(1 +\gamma)^2/[L +(1 + \alpha)^2(1 + \gamma)^2]$.

Much insight can be gained by analyzing the behavior of the system
in the phase plane (α , γ). The results of the phase plane analysis
(Goldbeter and Erneux, 1978; Goldbeter, 1980) which extend to a
larger number of protomers (Venieratos and Goldbeter, 1979), are
summarized in Figure 2a. Due to autocatalysis, the nullcline
$(d\gamma/dt) = 0$ possesses a region of negative slope. The steady state
(α_0, γ_0) which is located at the intersection of the two nullclines
is unstable whenever it lies in a region of sufficiently negative
slope of the sigmoid nullcline. Then the system evolves to a stable
limit cycle enclosing the unstable steady state (dashed curve in
Figure 2a). This regime corresponds to sustained oscillations in
the substrate and product concentrations (Figure 2b).

By increasing the substrate injection rate from a low initial
value, the $(d\alpha/dt) = 0$ nullcline can be shifted from left to
right so that the system passes from a stable steady state into
the oscillatory regime, and back to a stable steady state (Figure
2a). This accounts for the observation that glycolytic oscillations
in yeast extracts occur only when the substrate injection rate is
comprised between 20 and 160 mM/h (Hess et al, 1969).

The allosteric model for phosphofructokinase permits one to
explain how the oscillations break down upon addition of positive
or negative effectors which both decrease the cooperativity of the
enzyme below a critical level (Goldbeter and Venieratos, 1980).
The possible effect of fructose 2,6-bisphosphate on the oscillations
has been investigated in a recent theoretical study (Demongeot and
Kellershohn, 1983).

Excitability and oscillations at the cellular level : the cAMP signalling system of Dictyostelium discoideum

Experimental observations

The slime mold Dictyostelium discoideum represents a model
system whose study bears on many aspects of developmental biology
(Bonner, 1967; Loomis, 1982). D. discoideum amoebae grow as unicellular organisms and divide as long as food is available. Upon
starvation, they enter a period of interphase which lasts several
hours. Therafter they aggregate and form a migrating slug which
finally transforms into a fruiting body. D. discoideum provides
a model for studying both the transition between unicellular and
multicellular stages in the life cycle of a single organism, and

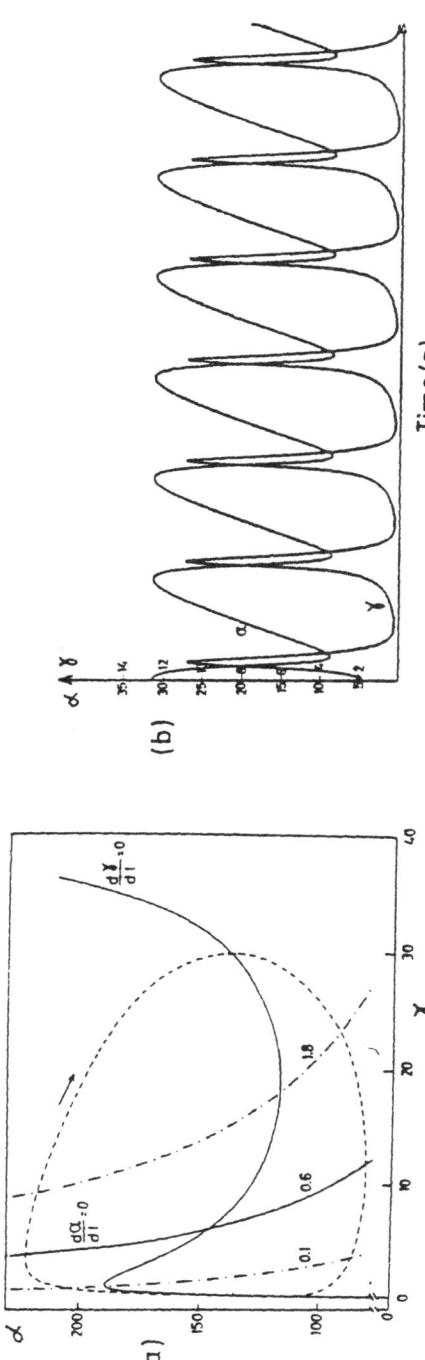

Fig.2. Phase plane behavior (a) and time evolution (b) of the product-activated reaction of Fig.1 governed by eqs. (1). The substrate and product concentrations are denoted by α and γ. In (a), the nullcline ($d\alpha/dt = 0$) is shown for three values of the normalized substrate input, v. When $v = 0.6 \ s^{-1}$, the steady state is unstable and the system evolves to a stable limit cycle (dashed line) corresponding to sustained oscillations of the type shown in (b).

the mechanism of differentiation, given that the amoebae finally
become either stalk cells or spores. We shall focus here on another
aspect of D. discoideum development which relates to the periodic
nature of the aggregation process.

The amoebae aggregate by a chemotactic response to cAMP signals
emitted by cells which behave as aggregation centers (Darmon and
Brachet, 1978; Gerisch, 1982). The centers release the cAMP signals
autonomously, with a periodicity of several minutes. As a result,
the aggregation proceeds in a wavelike manner (Gerisch, 1968; Alcantara and Monk, 1974). Movies of D. discoideum cells aggregating on
agar show concentric or spiral waves of amoebae moving towards the
centers (Gerisch, 1963). These patterns present a striking resemblen-
ce to the purely chemical waves observed in the Belousov-Zhabotinski
reaction (Winfree, 1974).

In D. discoideum , a center controls the aggregation of up to
10^5 amoebae. The aggregation territory can acquire dimensions which
are extremely large, of the order of a cm, as compared to the dimensions of single cells, i.e. 50 µ. The reason for the large range of
action of the aggregation centers was first envisaged by Shaffer
(1962) who hypothesized that cells responding chemotactically to the
signals emitted periodically by the centers can relay these signals.
Relay consists in the pulsatory amplification of a suprathreshold chemotactic signal. The nature of the chemical signal in D. discoideum
was later identified as being cAMP (Konijn et al, 1969). Other cellular slime molds may use other chemotactic signals. Some species lack
both the periodic release and the relay of the signal. In such species,
the chemotactic stimulus cannot be propagated over large distances,
and the size of the aggregation territory is consequently small. Hence
the name of Dictyostelium minutum given to one type of slime mold
that exemplifies this mode of aggregation (Gerisch, 1968).

Experiments carried out in cell suspensions have confirmed
the existence of oscillations (Gerisch and Hess, 1974; Gerisch and
Wick, 1975) and relay (Roos et al, 1975) of cAMP in D. discoideum.
Relay of suprathreshold cAMP pulses has also been observed in
experiments in which cells aggregating on agar were subjected to
artificial cAMP stimuli (Robertson and Drage, 1975).

Three main questions arise as to the cAMP signalling mechanism
in D. discoideum. First, why are the synthesis and release of
chemotactic signals periodic, whereas they are monotonic in other
slime mold species such as D. minutum ? Second, are relay and oscillations of cAMP due to two distinct mechanisms, as suggested by some
authors on the basis of experimental observations (Geller and
Brenner, 1978), or do they originate from a single biochemical mechanism ? A third question relates to the sequence of developmental transitions observed during interphase. Observations on agar

(Gingle and Robertson, 1976) and in cell suspensions (Gerisch et al, 1979) show that after starvation, the amoebae first acquire the capability of relaying cAMP stimuli, before being able to synthesize cAMP periodically, in an autonomous manner. The analysis of models for the cAMP signalling system, based on the regulatory properties observed in D. discoideum , provides unified answers to these issues.

Models for the cAMP signalling system : substrate depletion vs. receptor desensitization

One of the most conspicuous properties of cAMP production in D. discoideum is its character of self-amplification through positive feedback. The experimental observations on this system can be summarized as follows (Darmon and Brachet, 1978; Gerisch, 1982) (see Figure 3a). Cells possess membrane-bound receptors for cAMP, facing the extracellular medium; the cAMP receptor is functionally coupled to adenylate cyclase which transforms intracellular ATP into cAMP. Upon binding to the receptor, extracellular cAMP thus activates the synthesis of intracellular cAMP. The latter metabolite is transported into the extracellular medium where it is hydrolyzed by the membrane-bound and extracellular forms of phosphodiesterase; the role of this enzyme is to keep the extracellular cAMP at a low level so that cells can be receptive to successive signals arriving from the center.

The synthesis of cAMP in D. discoideum therefore possesses a definite autocatalytic nature. As for glycolysis, where the positive feedback exerted on phosphofructokinase gives rise to the instability associated with oscillations, we have suggested (Goldbeter and Segel, 1977; Goldbeter, 1975) that the positive feedback exerted on adenylate cyclase by extracellular cAMP is responsible for the periodic synthesis of cAMP in the slime mold. This regulatory mechanism is at the core of the models that we have analyzed for the cAMP signalling system (Figure 3a,b).

Besides cAMP, several factors such as H^+ and Ca^{++} ions may play a role in the mechanism of oscillations. Some experimental (Malchow et al, 1982; Europe-Finner et al, 1984) and theoretical (Rapp and Berridge, 1977) studies attribute to calcium a part in the control of cAMP oscillations. Thus pulses of a calcium ionophore induce a permanent phase-shift of the oscillations (Malchow et al, 1982), whereas sufficiently large amounts of EGTA inhibit aggregation (Europe-Finner et al, 1984). The approach that we have followed has been to concentrate on the globally autocatalytic nature of the cAMP signalling system, and to analyze its consequences for relay and oscillations. Experiments with adenosine, which inhibits cAMP binding, confirm that binding of extracellular cAMP to the receptor plays a primary role in the mechanism of oscillations

(Newell and Ross, 1982). As the detailed mechanism of activation
of adenylate cyclase which follows cAMP binding is still unclear,
we have analyzed successively two models based on the positive
feedback regulation (Figures 3a,b).

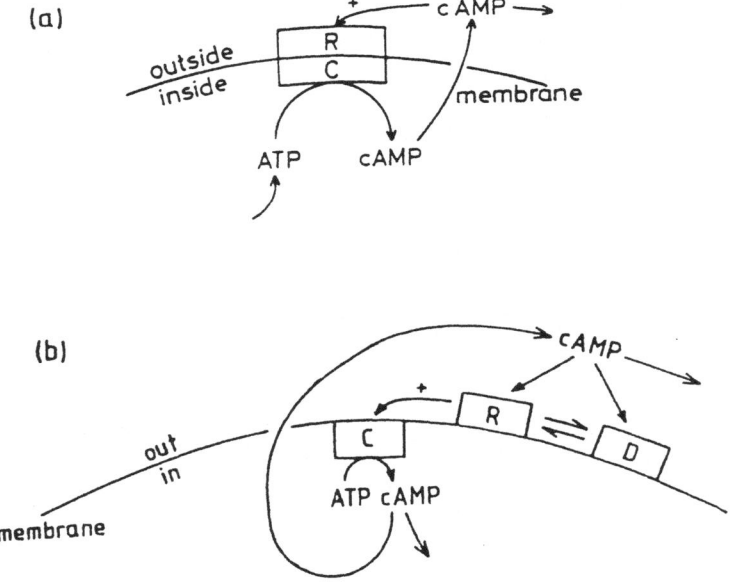

Fig. 3. Models for the cAMP signalling system of D. discoideum
based on coupling of the cAMP receptor (R) to adenylate cyclase
(C). In (b) , the receptor undergoes a reversible transition from
the active (R) to the desensitized (D) state of the receptor.

Periodic behavior always results from the interplay between two antagonizing effects. In glycolytic oscillations (Figure 2b), the rising phase of the product concentration results from autocatalysis, whereas the decreasing phase is due to the limiting effect of substrate depletion. Indeed, when the enzyme is activated by the product, it shifts from the T to the R state (See Figure 1) which consumes the substrate at an increased rate; since the substrate is supplied at a constant rate, eventually it begins to decrease – and so does the product – when the enzyme reaction rate exceeds the substrate input. In the model of Figure 3a, a similar mechanism for the limitation of cAMP autocatalysis through substrate depletion is envisaged. There we assume that the substrate ATP is produced at a constant rate, or that it is transported to the adenylate cyclase reaction site from a constant ATP pool. Other limiting effects may replace substrate depletion in the mechanism of oscillations. The most likely is probably receptor desensitization (Figure 3b). Before discussing in detail this second model, we shall recall the main results obtained for the model of Figure 3a.

Oscillations of cAMP

The model of Figure 3a comprises three variables, intracellular ATP, intracellular cAMP and extracellular cAMP whose reduced concentrations are denoted by α, β and γ, respectively. The time evolution of these variables is governed by a set of three ordinary differential equations when the effect of diffusion is neglected. The latter assumption is justified by the fact that we first wish to account for the behavior of D. discoideum in the cell suspension experiments (Gerisch and Hess, 1974; Gerisch and Wick, 1975; Roos et al, 1975). The nonlinearity of these kinetic equations, which is a prerequisite for periodic behavior (Prigogine, 1967; Nicolis and Prigogine, 1977) stems from the positive feedback exerted by cAMP and from the cooperative behavior of the receptor-cyclase system which is treated as a membrane-bound, allosteric complex with regulatory (receptor) sites facing the extracellular medium, and catalytic sites facing the inside of the cell (Goldbeter and Segel, 1977) (the assumption of a close association between receptor and cyclase is relaxed in the model of Figure 3b considered below).

The set of kinetic equations governing the model can be reduced to a system of two differential equations for α and γ by a quasi-steady-state hypothesis for β (Goldbeter et al, 1978). Then the system of kinetic equations reduces to a form similar to equs. (1) and the phase portrait presents the same aspect as that shown in Figure 2a (Goldbeter, 1980; Goldbeter et al, 1978). Namely, the steady state is unstable when located on a region of sufficiently negative slope on the nullcline $(d\gamma/dt)=0$. Sustained oscillations in α(ATP) and γ(extracellular cAMP) develop in these conditions.

This behavior accounts for the periodic synthesis of cAMP observed
in cell suspensions, as well as for the periodic release of cAMP
pulses by aggregation centers in populations of starving amoebae on
agar.

Relay of cAMP pulses : excitability

The phenomenon of relay occurs in closely related conditions
(Figure 4a). Here, the steady state is located in the immediate
vicinity of the oscillatory domain, on the left of the region of
negative slope on the sigmoid nullcline. The steady state (α_o, γ_o)
is now stable. We wish to determine the response of the system
to a pulse of cAMP. Such a perturbation amounts to a horizontal
displacement of the system away from the steady state since, initially, the ATP level remains unchanged whereas extracellular cAMP has
been increased instantaneously. In curve (a) of Figure 4a, a small
displacement corresponding to the application of a small cAMP pulse
results in the immediate return of the system to the stable steady
state : the perturbation has been damped (see the corresponding
curve in Figure 4b). Above a threshold, however, the perturbation is
amplified in a pulsatory manner as the system undergoes a large
excursion in the phase plane before returning to the stable steady
state (curves (b) in Figures 4a,b). Such behavior reflects the excitability of the cAMP signalling system. This property can be identified with the relay of cAMP pulses which has been demonstrated in
cells aggregating on agar as well as in suspensions.

Both the existence of a sharp threshold and of a refractory
period for relay (Robertson and Drage, 1975) can be explained by
the diagram of Figure 4a. The refractory period is the time required
to return to the steady state after the large excursion associated
with relay. Only when the system is close enough from this state,
can it be excited by a second stimulus. The sharp threshold suggested by Figure 4 becomes more apparent in a dose-response curve
which shows the amplitude of the cAMP response as a function of the
magnitude of the cAMP pulse (Goldbeter and Segel, 1977; Goldbeter,
1980; Goldbeter et al, 1978).

The phase plane analysis of Figures 2 and 4 explains why excitability and autonomous oscillations are necessarily associated
within the same system : both phenomena represent two different
modes of dynamic behavior which occur for closely related parameter
values. Hence there is no need for two distinct mechanisms to account
for relay and oscillations of cAMP. A shift in parameter values can
explain why cells exhibit either one (or none) of these properties
during development, and how they may switch from oscillations to
excitability upon addition of certain chemicals (Geller and Brenner,
1978).

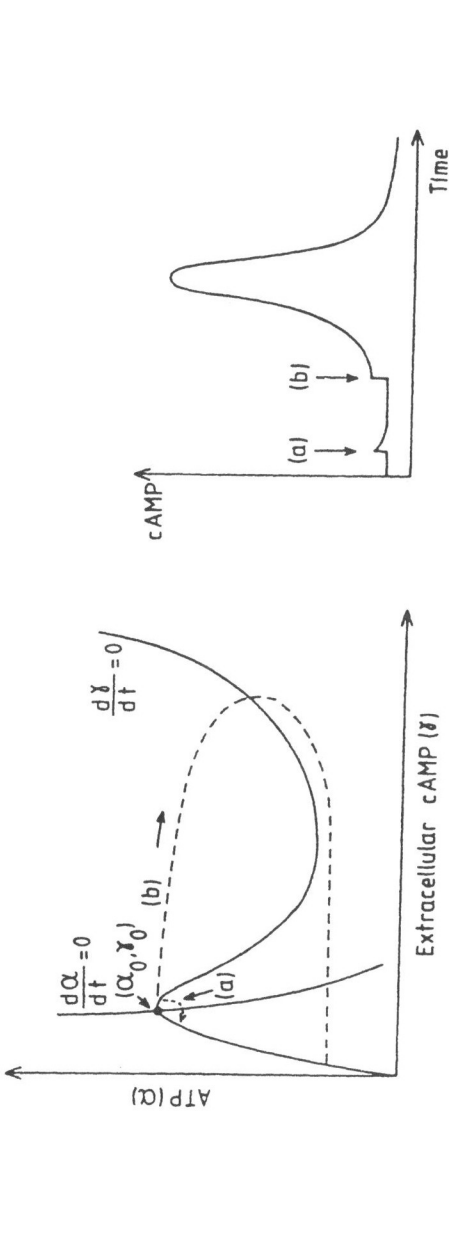

Fig. 4. Excitability in D. discoideum . The response to subthreshold (a) and suprathreshold (b) cAMP pulses is shown in the phase plane and as a function of time.

Such a close association between excitability and oscillations
is also observed in chemical systems(Winfree , 1972; de Kepper,1976).
The link between the two phenomena has been stressed long ago by
neurophysiologists, and was particularly emphasized in the book by
Fessard(1936). Later on, in his theoretical study of the nerve
membrane, Fitzhugh (1961) introduced phase plane analysis to show
how oscillations and excitability occur for closely related parameter values in the same model. Although the molecular mechanism of
oscillations in the two cell types differ, D. discoideum amoebae
and nerve cells share common dynamic properties. In two of his
papers Durston (1973, 1974) emphasized this relationship by refer-
ring to the excitable and pacemaker properties of D. discoideum
cells.

Developmental control of the cAMP signalling system

The analysis of a unified mechanism for cAMP relay and oscillation provides an explanation for the sequence of developmental transitions observed in the signalling system after starvation (Goldbeter and Segel, 1980). Such analysis also provides a physico-chemical
basis for the difference between centers and relay cells. The likelihood of genetic differences between centers and relay cells is
indeed very small, as experiments indicate that one cell out of five
can become an aggregation center (Glazer and Newell, 1981).

Up to two hours after the beginning of starvation, cells are
unable to relay cAMP signals. Then they acquire this property, and
some two hours later they become capable of autonomous oscillations
of cAMP. After a few hours, cells lose the potential for spontaneous
oscillations and become excitable again. These observations on the
development of the signalling properties in cell suspension
(Gerisch et al, 1979) corroborate those made for cells aggregating
on agar (Robertson and Drage, 1975; Gingle and Robertson, 1976).

To examine the mechanism underlying the sequence of developmental transitions, it is useful to determine the behavior of the model
as a function of two key parameters of the signalling system, i.e.
adenylate cyclase and phosphodiesterase. This can be done by linear
stability analysis and by numerical simulations. The former method
consists in determining the evolution of infinitesimal perturbations
around the steady state (Nicolis and Prigogine, 1977). If these
perturbations grow in time, the steady state is unstable and the
system either undergoes sustained oscillations around the unstable
state (as in Figure 2) or evolves towards another stable steady
state when the kinetic equations admit such multiple solutions.

The behavior of the signalling system as a function of the
maximum activity of adenylate cyclase and of phosphodiesterase is
illustrated in Figure 5. In the dashed area C, the unique steady

state   admitted by the system is unstable. Spontaneous oscillations
of cAMP occur in this domain. In the relay domain B, the unique
steady state is stable but excitable; the system amplifies in a
pulsatory manner a cAMP pulse of given magnitude. In E, the system
admits three steady states, two of which can be stable. Outside these
domains, in A, the signalling system admits a unique steady state
which is stable but not excitable.

    At the beginning of starvation, the levels of phosphodiestera-
se and adenylate cyclase in D. discoideum are low. In the diagram
of Figure 5, cells would thus start the interphase near A. In order

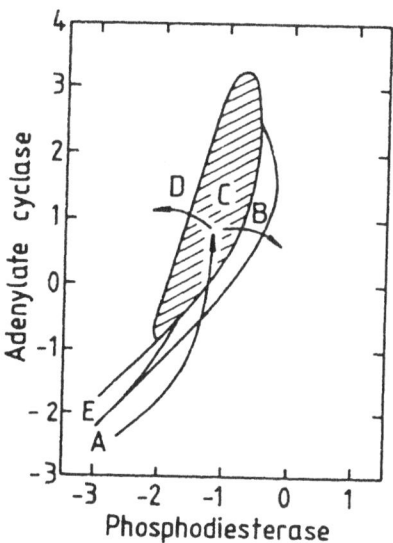

Fig. 5. Developmental path for the cAMP signalling system of D. discoideum . The diagram, established in logarithmic scale, shows the
domains of relay (B) and oscillations (C). The path symbolized by
the arrow from A accounts for the sequence of transitions observed
after starvation (Goldbeter and Segel, 1980).

to account for the sequence no relay-relay-oscillations, the amoebae would have to follow a developmental path represented by the arrow passing successively through regions A, B and C. This path corresponds to an increase in adenylate cyclase and phosphodiesterase activity. Such increase in the two enzyme activities is indeed observed after starvation (Klein and Darmon, 1977; Klein, 1976; Coukell and Chan, 1980). Thus a continuous variation in enzyme activity could underlie the sequence of discontinuous transitions observed in the dynamic properties of the cAMP signalling system.

A detailed study of D. discoideum mutants would permit an experimental verification of the above suggestions (Goldbeter and Segel, 1980). In support of the theoretical predictions is the observation (Coukell and Chan, 1980) that an early rise in adenylate cyclase is associated with a precocious onset of cAMP oscillations in the rapid developing mutant FR17. On the other hand, a mutant lacking phosphodiesterase becomes able to aggregate in a periodic manner when supplied with this enzyme (Darmon et al, 1978).

As to the nature of aggregation centers, Figure 5 suggests that they are the cells which are the first to enter the oscillatory domain C after starvation (Goldbeter, 1975; Goldbeter and Segel, 1980). As soon as their content in adenylate cyclase and phosphodiesterase is such that the two enzyme activities correspond to a point located in C, spontaneous oscillations of cAMP set in, and they begin to release periodically pulses of cAMP. Numerical simulations show that there is practically no delay for the onset of large-amplitude oscillations as the system passes through the bifurcation point.

It is natural that in a population of some 10^5 amoebae, some cells are more advanced than others on the developmental path and therefore are the first to enter the instability domain of the adenylate cyclase reaction. If these cells are removed, other cells further away on the path - e.g., in the relay domain B - will take their place as they enter in turn the oscillatory mode. The fact that cells lose the capability of oscillations after a few hours could be due to a second wave of phosphodiesterase synthesis (Klein and Darmon, 1977) which, as indicated by an arrow in Figure 5, would bring the cells back into the relay domain. This would also account for the late transition from oscillations to relay observed in cell suspensions (Gerisch et al, 1979), and for the eventual disappearance of new centers in cell populations aggregating on agar .

The developmental path approach proposed for D. discoideum is of general significance for the development of biological systems in which it could account for the onset of spatial (Kauffman et al, 1978) or temporal organization. The continuous elevation

(or decrease) in some relevant biochemical parameters (enzyme activities, receptor levels, etc...) could induce the passage through
bifurcation points separating qualitatively different types of behavior. Such phenomenon leading to the entrance into a domain of limit
cycle oscillations most probably underlies the onset of beating in
developing heart cells (de Haan, 1980) , as well as the onset of
hormonal rhythms at puberty (Smith, 1983).

Model for the cAMP signalling system based on receptor desensitiza-
tion

     Two sets of experiments appear to rule out substrate depletion
as the main cause underlying the decreasing phase of cAMP synthesis
during relay and oscillations. Measurements of ATP levels during
cAMP oscillations show only a very slight variation of intracellular
ATP (Roos et al, 1977). Compartmentation of ATP within the cell can,
however, not be excluded (Roos et al, 1977). A second type of experiments in which the cells are subjected to constant cAMP stimuli
(see section below) showed that a second response can be elicited
by an increase in the extracellular cAMP stimulus; this shows that
cAMP synthesis is not limited by ATP depletion (Devreotes and Steck,
1979).

     In looking for a limiting effect that would substitute for
substrate depletion in balancing the autocatalysis exerted by
cAMP, we reached the conclusion that receptor desensitization is
a most likely candidate for such function (Goldbeter and Martiel,
1980). This phenomenon represents the passage of the active receptor – i.e. the state of the receptor which, upon binding of cAMP,
produces the activation of adenylate cyclase – into an active state
uncoupled from the enzyme. The inactive or desensitized state D
(see Figure 3b) represents either another conformational state,
as in the model proposed by Katz and Thesleff (1957) for the acetylcholine receptor, or a covalently modified form of the receptor.

     Experimental evidence for receptor desensitization has been
obtained for neurotransmitters such as acetylcholine (Heidmann and
Changeux, 1978), and for a variety of hormonal systems. The phenomenon may involve covalent modification such as phosphorylation in
the case of the β-adrenergic receptor (Stadel et al, 1983), or
methylation which is associated with adaptation to constant stimuli
in bacterial chemotaxis (Koshland , 1979; Springer et al, 1979).
In D. discoideum , Klein has shown that the receptor exists in
two states possessing, respectively, a high and a low affinity of the
cAMP (Klein, 1979). These could represent the R and D states of the
receptor in Figure 3b. Moreover, a membrane protein having a
molecular weight close to that of the cAMP receptor is phosphoryla-
ted after exposure to cAMP (Lubs-Haukeness and Klein , 1982).

Protein phosphorylation could thus result in the uncoupling of the receptor from adenylate cyclase (Lubs–Haukeness and Klein, 1982). Recent experiments (Theibert and Devreotes, 1983) also suggest that receptor modificiation takes place during adaptation of the cAMP signalling system to constant stimuli.

The following treatment of the model of Figure 3b applies equally to the reversible transformation R $\rightleftharpoons$ D viewed as a transition between two conformational states, or as a process of covalent modification. We assume that only the R state is coupled to adenylate cyclase. Extracellular cAMP binds to the R and D states with different affinities. To ensure positive cooperativity in binding (Coukell, 1981), we assume that two molecules of cAMP bind to the receptor. The cAMP–receptor complex activates adenylate cyclase through a simple collision–coupling mechanism (Tolkovsky et al, 1982). The reaction steps of the model are detailed as follows :

$$R \underset{k_{-1}}{\overset{k_1}{\rightleftharpoons}} D$$

$$R + 2P \underset{d_1}{\overset{a_1}{\rightleftharpoons}} RP_2 \quad, \quad D + 2P \underset{d_2}{\overset{a_2}{\rightleftharpoons}} DP_2$$

$$RP_2 \underset{k_{-2}}{\overset{k_2}{\rightleftharpoons}} DP_2 \qquad\qquad\qquad (2)$$

$$RP_2 + C \underset{d_3}{\overset{a_3}{\rightleftharpoons}} E$$

$$E + S \underset{d_4}{\overset{a_4}{\rightleftharpoons}} ES \overset{k_4}{\longrightarrow} E + P_i$$

$$C + S \underset{d_5}{\overset{a_5}{\rightleftharpoons}} CS \overset{k_5}{\longrightarrow} C + P_i$$

$$P_i \overset{k_i}{\longrightarrow} \quad, \quad P_i \overset{k_t}{\longrightarrow} P \quad, \quad P \overset{k_e}{\longrightarrow}$$

where P_i and P denote intra- and extracellular cAMP, respectively; S is ATP; C is the free form of adenylate cyclase which is taken as less active than in the complex E formed with the cAMP receptor; k_i and k_e are the apparent first-order rate constants of the intracellular and extracellular forms of phosphodiesterase (Dinauer et al, 1980) , whereas k_t measures the linear rate of transport of cAMP into the extracellular medium (Dinauer et al, 1980). The time evolution of the model is governed by the following set of kinetic equations:

$$\frac{d\rho}{dt} = k_1(-\rho + L_1\delta) + d_1(-\rho\gamma^2 + x)$$

$$\frac{d\delta}{dt} = k_1(\rho - L_1\delta) + d_2\left[1 - \delta(c^2\gamma^2 +1)-\rho-x-\mu(1-\bar{c}(1+\alpha\theta))\right]$$

$$\frac{dx}{dt} = k_2\left[L_2(1-\rho-\delta)-(1+L_2)x - L_2\mu(1-\bar{c}(1+\alpha\theta))\right] +d_1(\rho\gamma^2-x)+$$

(3)
$$+\mu d_3\left[\phi -\bar{c}(\epsilon x +\phi(1+\alpha\theta))\right]$$

$$\frac{d\bar{c}}{dt} = d_3\left[\phi -\bar{c}(\epsilon x +\phi(1+\alpha\theta))\right]$$

$$\frac{d\alpha}{dt} = v -\sigma\alpha\phi\left[1-\bar{c}(1 + \alpha\theta(1-\lambda) - \lambda\theta)\right]$$

$$\frac{d\beta}{dt} = q\sigma\alpha\phi\left[1-\bar{c}(1+ \alpha\theta(1-\lambda) -\lambda\theta)\right] - (k_i+k_t)\beta$$

$$\frac{d\gamma}{dt} = (k_t\beta/h) - k_e\gamma+ 2\eta\left[d_1(-\rho\gamma^2+x)+d_2(1-\rho-\delta(1+c^2\gamma^2)-\right.$$

$$\left.-x-\mu(1-\bar{c}(1+\alpha\theta))\right]$$

These equations have been obtained by taking into account the conservation relations for the total amounts of adenylate cyclase, E_T, and receptor , R_T, and by making a quasi-steady-state hypothesis for the complexes CS and ES. The reduced variables are defined as ρ = R/R_T, δ =D/R_T, x=RP_2/R_T, $\bar{c}$ = C/E_T, α =S/K_m with K_m=$(d_4+k_4)/a_4$, β =$P_i/K_R^{1/2}$ and γ= $P/K_R^{1/2}$, with K_R = d_1/a_1. Finally the parameters appearing in equations (3) are defined as L_1= k_{-1}/k_1, L_2 = k_{-2}/k_2,

$c = (K_D/K_R)^{1/2}$, $K_D = d_2/a_2$, $\theta = K_m/K_m'$ with $K_m' = (d_5 + k_5)/a_5$,
$\lambda = k_5/k_4$, $\varepsilon = R_T/K_E$ with $K_E = d_3/a_3$, $\sigma = k_5 E_T/K_m$,
$\phi = 1/(1+\alpha)$, $q = K_m/K_R^{1/2}$, $\eta = R_T/K_R^{1/2}$, $\mu = E_T/R_T$. Moreover, v represents
the constant rate synthesis of ATP, divided by K_m , and h is a
dilution factor.

Due to the large number of variables, the model is no more
amenable to a simple phase plane analysis of the kind performed in
Figures 2 and 4, unless it is reduced to a smaller number of variables under restrictive conditions (Goldbeter and Martiel, 1983).
Such reduction is made difficult by the fact that some kinetic equations contain a mixture of terms related to fast binding processes
and to the slower reactions of receptor modification. One can still
determine, however, the dynamic properties of the signalling system
by numerical integration of the differential equs. (3), preceded
by a linear stability analysis of the steady-state solution admitted by these equations.

The basic results obtained in the model of Figure 3a, namely
relay and oscillations of cAMP, are recovered in the model based on
receptor modification. The oscillations of cAMP are now accompanied
by a periodic evolution in the various forms of the receptor. The
phase relationship between the fraction of free, active receptor
and cAMP shown in Figure 6 indicates that the decreasing phase of
cAMP is due to a depletion of the pool of active receptor as the
latter is pumped into the desensitized state. Figure 6 also shows
that the cAMP synthesis can proceed periodically with only a slight
periodic variation in the level of ATP, as observed in the cell suspension experiments (Roos et al, 1977).

As in Figure 4, relay of cAMP pulses occurs in a parameter
domain close to that producing oscillations. The dose-response
curve for this excitable behavior is characterized by a sharp
threshold, as in the model of Figure 3a (Goldbeter and Segel, 1977;
Goldbeter et al, 1978).

Adaptation to constant cAMP stimuli

Aggregating D. discoideum amoebae respond to cAMP pulses released periodically by the centers. Between two pulses, the cAMP
signals are hydrolyzed by phosphodiesterase. Under natural conditions, extracellular cAMP is therefore subjected to sharp, periodic
variations. Bypassing these conditions, Devreotes et al (Devreotes
and Steck, 1979; Theibert and Devreotes, 1983; Dinauer et al,
1980) have carried out a series of experiments using constant cAMP
stimuli. Although clamping the extracellular cAMP levels changes
the dynamic properties of the system, these experiments have proved
to be extremely fruitful in revealing new properties of the signalling mechanism. The main result of such studies is that cells adapt
to constant stimuli (Devreotes and Steck, 1979) . In response to a

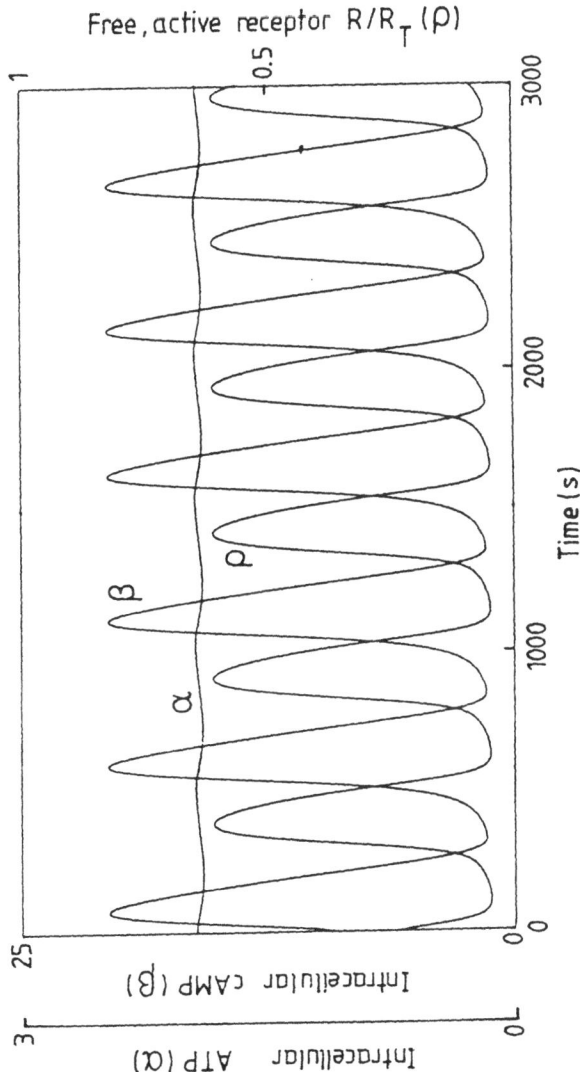

Fig. 6. Oscillations of cAMP (β) in the model for the cAMP signalling system based on receptor modification. Also shown are the periodic variation of ATP (α) and of the fraction of free active receptor, $\rho = R/R_T$. The curves are obtained by numerical integration of equs. (3) for the parameter values $\theta = \lambda = 0.01$, $\varepsilon = 0.2$, $\mu = \eta = 0.1$, $h = 6$, $L_1 = 316$, $k_2 = k_{-2} = 0$, $c = 100$, $k_1 = 0.0168$ s^{-1}, $q = 4000$, $v = 0.00022$ s^{-1}, $\sigma = 0.0105$ s^{-1}, $k_e = 0.084$ s^{-1}, $k_i = 0.063$ s^{-1}, $k_t = 0.042$ s^{-1}, $d_1 = d_2 = d_3 = 1$ s^{-1}, $K_R^{1/2} = 10^{-7}$ M.

step increase in extracellular cAMP, cells synthesize a pulse of
intracellular cAMP before the latter returns to prestimulus   level
after a few minutes, although extracellular cAMP is now maintained
at a higher level. Adaptation has also been demonstrated in similar
conditions for the cyclic GMP response (Van Haastert and Van der
Heijden, 1983) which follows stimulation by cAMP in D. discoideum .
The phenomenon of adaptation is a general property of sensory systems (Koshland et al, 1982), and is best characterized at the biochemical level in terms of receptor methylation in bacterial chemotaxis (Koshland, 1979; Springer et al, 1979). A model based
on receptor modification has been developed to account for adaptation
in the latter system (Goldbeter and Koshland , 1982).

Recent experiments of Theibert and Devreotes (1983) have shown
that adaptation can occur in D. discoideum in the absence of adenylate
cyclase activation. Using caffeine, which uncouples the cAMP   receptor from adenylate cyclase, they showed that incubation of cells with
caffeine and cAMP precludes the synthesis of a cAMP pulse even when the
uncoupling   agent is removed after a few minutes. These and other
data suggest that the process of adaptation involves the receptor
rather than adenylate cyclase (Theibert and Devreotes, 1983).

The model of Figure 3b, detailed by the sequence of steps (2),
accounts for these results. In Figure 7 is shown the response of the
model when the level of extracellular cAMP, i.e. P in   equs (3), is
raised from 0 to 10^{-7} M (a) or 10^{-6}M(b). Now P is a parameter in
equs. (3), since it is held constant as in the   clamp experiments.
Figure 7 shows that the signalling system (2) adapts to step increases in extracellular cAMP. As in the experiments (Devreotes and
Steck, 1979), both the duration and the amplitude of the response increase with the magnitude of the stimulus. The fact that the receptor modification model can account for this experimental observation
shows that the doubts which have been raised in this respect
(Devreotes and Steck, 1979) are ill-founded. In fact, the relative
duration of the response to constant stimuli of various magnitudes
can be tuned mainly by varying constant k_{-1} which in the model
governs the return of the receptor from the desensitized to the
active state.

Curve (c) in Figure 7 shows the simulation of a caffeine experiment (Theibert and Devreotes, 1983). There, the system is
subjected at time zero to a step increase $0 \longrightarrow 10^{-6}$ M in extracellular cAMP, but the coupling   between the receptor and adenylate cyclase is prevented during 300 s. After this period, cells fail
to respond to the cAMP stimulus, although the coupling has been
restored. As in experiments (Theibert and Devreotes, 1983), adaptation has occurred without cAMP synthesis, owing to receptor modification.

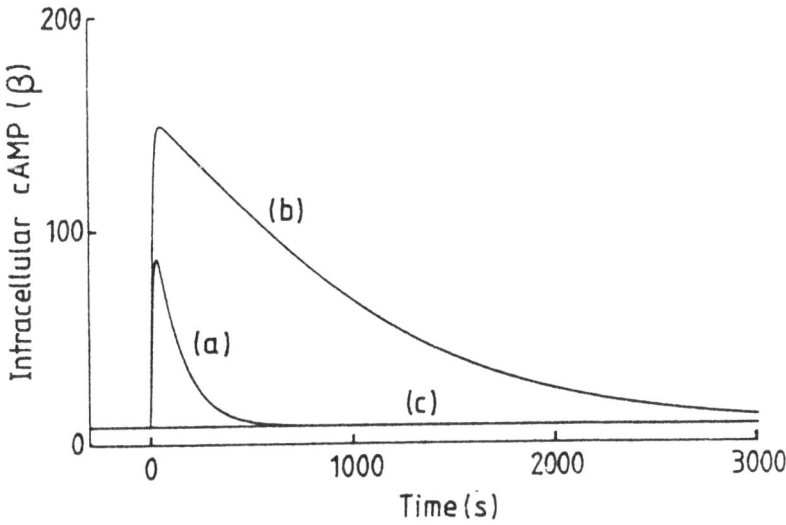

Fig. 7. Adaptation of the cAMP signalling system to constant stimuli. The model based on receptor modification (equs. (2) and (3)) is subjected at time zero to an increase in extracellular cAMP from 0 to 10^{-7} M in (a), and to 10^{-6}M in (b). The resulting time evolution of intracellular cAMP is obtained by numerical integration of equs. (3) in which extracellular cAMP (γ) is held constant at a value corresponding to 10^{-7} or 10^{-6}M. In (c) , no response is observed for the stimulus of (b) when the receptor is uncoupled from adenylate cyclase (by making $\varepsilon = 0$ in equs. (3)) during the first 300 s. Parameters values are those of Fig. 6.

The model further explains why there is no observable threshold when relay is determined in response to step increases in extracellular cAMP (Devreotes and Steck, 1979). The reason beomes apparent from Figure 8 . In (a) is shown the classical picture where binding of a hormone to a specific cell-surface receptor elicits the synthesis of the second messenger cAMP, through coupling between the receptor and adenylate cyclase. In (b), the particularity of D. discoideum is that cAMP behaves as the extracellular hormone signal – i.e. as the first, and as the second messenger (Konijn , 1972) ! This results in a positive feedback loop which produces the instability associated with autonomous oscillations, and the excitability iden- tified with relay of suprathreshold cAMP pulses.

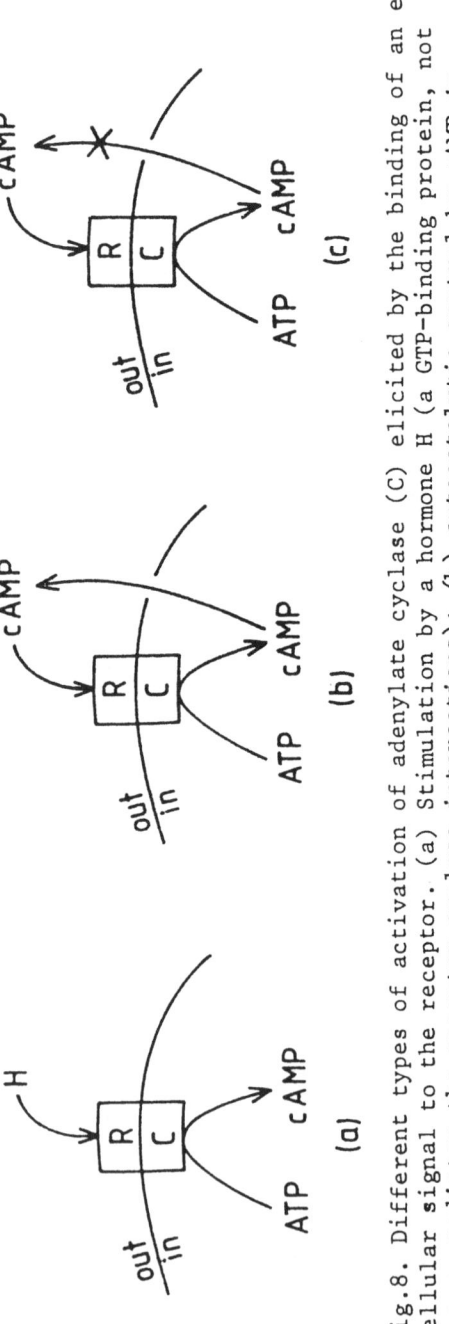

Fig.8. Different types of activation of adenylate cyclase (C) elicited by the binding of an extra-cellular signal to the receptor. (a) Stimulation by a hormone H (a GTP-binding protein, not shown, mediates the receptor-cyclase interactions); (b) autocatalytic control by cAMP in D. discoideum; (c) the positive feedback of (b) is suppressed when the external cAMP is maintained at a constant level.

In (c), the control of the extracellular cAMP level in clamp
experiments signifies that the positive feedback loop is suppressed
in these conditions. Indeed, at the receptor, the system "ignores"
that it further produces cAMP since the stimulus is maintained
constant. In the absence of self-amplification , the signalling system becomes comparable to a regular hormone signalling mechanism
such as illustrated in (a). No oscillations occur in these (artificial) conditions. Relay can occur, as shown in Figure 7, but there
is no more a sharp threshold in the dose-response curve. Such threshold is indeed a characteristic of excitable systems, and excitability requires self-amplification through positive feedback.

COMPLEX OSCILLATIONS, BIRHYTHMICITY AND CHAOS

We now quit the cAMP signalling system of D. discoideum which
illustrates excitability and periodic behavior, and turn to the
question of how complex oscillatory phenomena arise in regulated
biochemical systems. At the end of this section, we shall return
to the cAMP signalling system and will show how complex phenomena
also occur, rather unexpectedly, in the model of Figure 3b.

Phenomena resulting from the interplay between two instability-
generating mechanims

Until now we have examined the different types of dynamic behavior encountered in systems controlled by a single positive feedback process. The most common types of nonequilibrium phenomena
associated with a single instability-generating mechanism are
sustained oscillations around the unstable steady state, and excitability when the steady state is stable but close to the domain of
oscillations. Another common situation is that of two stable steady
states which coexist for the same set of parameter values. This
phenomenon, referred to as bistability, has been demonstrated experimentally in many systems, including some biochemical reactions
(Degn, 1968; Naparstek et al, 1974; Eschrich et al, 1980).

The question arises as to whether additional modes of dynamic
behavior appear when the regulatory structure of a chemical network
comprises more than one instability-generating mechanism. The
analysis of a model for a multiply regulated biochemical system
shows that the variety of behavioral modes is then greatly enhanced
(Decroly and Goldbeter, 1982; Goldbeter and Decroly, 1983).

The model considered (Figure 9) is that of a sequence of two
allosteric enzymes, E_1 and E_2, which transform respectively substrate S into product P_1, and P_1 into product P_2. Each enzyme is
activated by its reaction product. The first substrate is injected
at a constant rate v, whereas the final product is removed at a
rate proportional to its concentration, with an apparent first

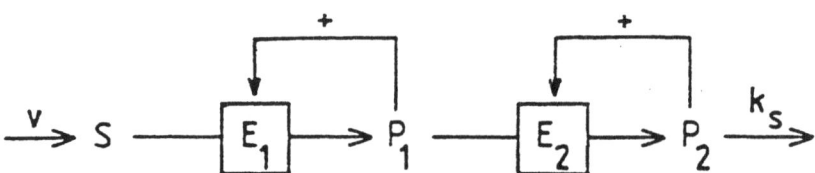

Fig.9. Model of two autocatalytic enzyme reactions coupled in series.

order rate constant k_s. The model thus comprises three variables
- i.e. the reduced concentrations of S, P_1 and P_2 denoted,respec-
tively , by α, β and γ - whose time evolution is governed by the
following set of nonlinear differential equations :

$$\frac{d\alpha}{dt} = (v/K_{m1}) - \sigma_1 \phi$$

$$(4) \qquad \frac{d\beta}{dt} = q_1\sigma_1\phi - \sigma_2\eta$$

$$\frac{d\gamma}{dt} = q_2\sigma_2\eta - k_s\gamma$$

with $\phi = \alpha(1+\alpha)(1+\beta)^2 / [L_1 +(1+\alpha)^2(1+\beta)^2]$

$\eta = \beta(1+d\beta)(1+\gamma)^2 / [L_2 + (1+d\beta)^2(1+\gamma)^2]$

The detailed analysis of these equations has been carried out in
two recent publications (Decroly and Goldbeter, 1982; Goldbeter
and Decroly, 1983) where further information can be found. Here,
we recall only the salient features of this analysis.

The behavior of the model was primarily studied as a function
of parameters v and k_s. Upon changing continuously the value of
k_s, a variety of complex oscillatory phenomena can be observed.
They can be visualized on a schematic bifurcation diagram showing
the steady-state level of substrate S (α_0) and the maximum amplitu-

de of S in the course of oscillations (α_M) as a function of k_s (Figure 10). Stable and unstable regimes are represented by solid and dashed lines, respectively. The diagram is established by combination of linear stability analysis for the steady state, and numerical simulations for the stable and unstable periodic regimes (Decroly and Goldbeter, 1982). Four different types of dynamic behavior are illustrated in Figure 11 as a function of time, for increasing values of k_s.

It is clear from the preceding section that each of the feedback processes linking S to P_1 and P_1 to P_2 can give rise by itself to sustained oscillations. When constant k_s tends to zero, P_2

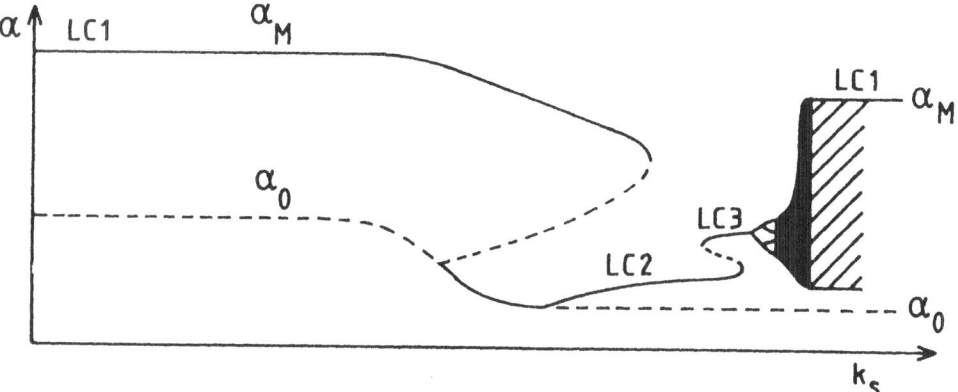

Fig. 10. Schematic bifurcation diagram for the model of Figure 9. The steady-state level of substrate (α_θ) and the maximum amplitude of substrate oscillations (α_M) are shown as a function of k_s. The various domains of the diagram are not drawn to scale, for the sake of clarity (see Fig. 12).

accumulates, whereas P_2 decreases to zero when k_s goes to infinity. In both cases, v can be chosen so that the first instability mechanism gives rise to simple periodic oscillations in S and P_1. Let us denote the corresponding stable limit cycle by LC1 (Figures 10 and 11a). When k_s is increased from a low initial value, the steady state becomes stable and coexists with limit cycle LC1. This gives rise to a phenomenon of hard excitation : small perturbations away from the stable steady state regress; however, when the perturbation exceeds a threshold, the system quits the steady rate and evolves to a stable periodic regime (Figure 11b).

For higher values of k_s, the steady state becomes unstable again and a new stable limit cycle, LC2, arises through a Hopf bifurcation. As LC1 at first persists, we observe a phenomenon of birhythmicity in which the two stable limit cycles LC1 and LC2 coexist, separated by an unstable periodic trajectory. Depending on initial conditions, the system will evolve to either one of the two stable oscillatory regimes which are characterized by distinct amplitudes and frequencies. To our knowledge, the phenomenon has not yet been observed in biological or biochemical systems. It has recently been demonstrated experimentally in a chlorite-bromate-iodide system(Alamgir and Epstein, 1983) which possesses a feedback structure analogous to that of Figure 9 as it comprises two coupled autocatalytic reactions. When k_s further increases in Figure 10, LC1 disappears. Then a hysteresis loop forms between LC2 and a new stable limit cycle LC3. This gives rise to a second birhythmic pattern.

Upon further increasing k_s, the LC3 oscillations undergo a sequence of period-doubling bifurcations. As in the sequence described by Feigenbaum(1978), limit cycles of period 1, 2, 4, 8,... are observed until a value of k_s is reached beyond which the oscillations cease to be periodic (Decroly and Goldbeter, 1982). Such aperiodic oscillations are referred to as chaos (Vidal and Pacault, 1981) (Figure 11c). They occur in the domain represented as a black area in Figure 10. Although governed by deterministic laws, chaotic behavior contains elements of both randomness and periodicity. Indeed, the trajectory followed by the system in the phase space never passes twice through any given point, although the system is attracted in a well defined portion of this space . Hence the name of strange attractor (Vidal and Pacault, 1981) given to such a trajectory.

In contrast to such behavior, complex periodic oscillations which are observed for higher values of k_s do not possess the unpredictability of chaos. They correspond to a complex trajectory in the phase space, but this trajectory is closed and repeats itself after a well-defined period. The term "folded limit cycle" has been proposed for such trajectories by Schulmeister and Sel'kov (1978) who observed them in a biochemical model for which they also

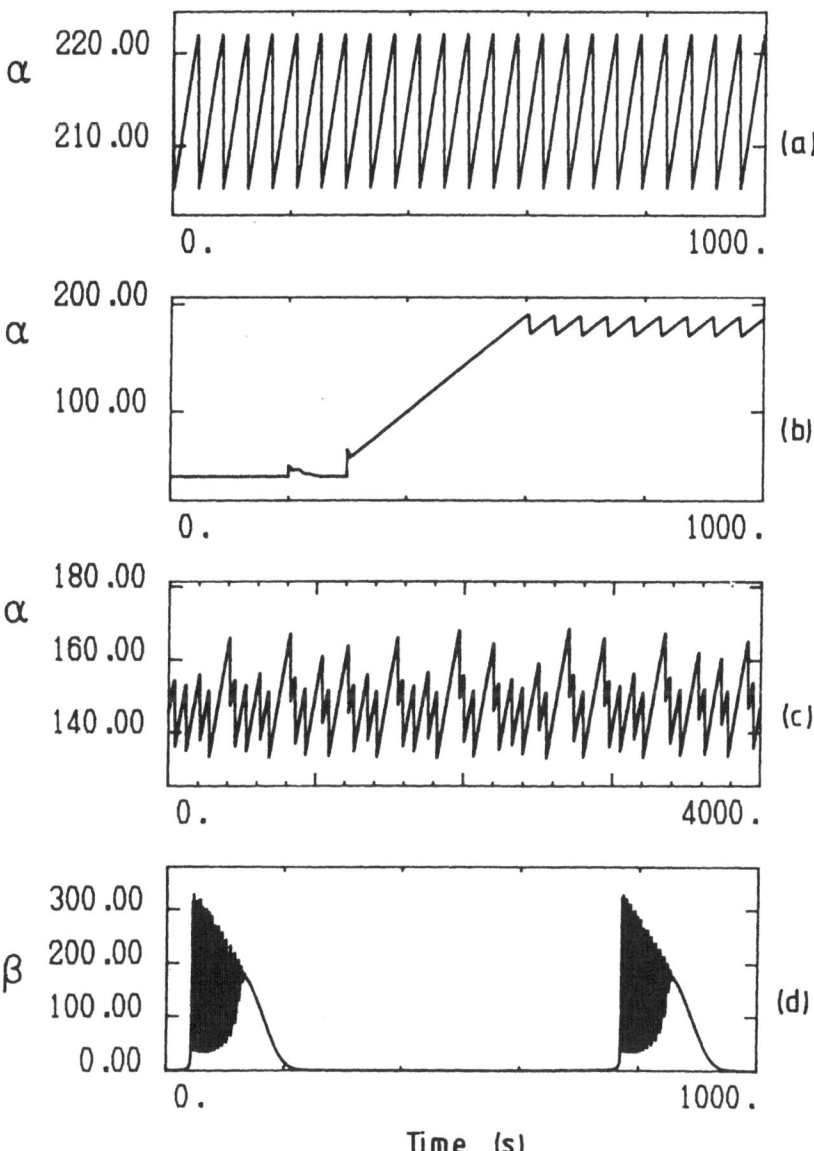

Time (s)

Fig.11. Time evolution in the model of Figure 9. The curves are obtained by integration of equs. (4) for (a) k_s =0.6 s^{-1}, (b) k_s=1.2 s^{-1}, (c) k_s =1.4 s^{-1}, (d) k_s=7 s^{-1}. Other parameters are (v/K_{m1}) =0.45 s^{-1}, L_1 = 5.10^8, L_2 = 100, σ_1= σ_2= 10 s^{-1}, q_1 = 50, q_2 = 0.02, d=0; v = 0.25 s^{-1} in (c) and (d) , and in the latter situation, q_2 = 0.1.

reported the coexistence between two stable periodic regimes. Complex oscillations in the present model correspond to a bursting phenomenon, as several sharp peaks in P_1 (β) and P_2 (γ) are observed over a period (Figure 11d). The waveform obtained in these conditions presents a striking resemblance with certain firing patterns of neurons.

To determine the relative importance of these various modes of oscillations, we have analyzed the dynamic behavior of the multiply regulated system as a function of parameters v and k_s (Figure 12). Simple periodic oscillations remain the most common mode of temporal self-organization, followed by complex periodic oscillations. Two domains of birhythmicity and three domains of chaos are observed, but these are small in comparison with the domains of simple or complex periodic behavior.

The two domains of birhythmicity shown in Figure 12 correspond to the existence of LC1 with LC2, or LC2 with LC3. In a small range of parameter values, these two domains overlap : this gives rise to a phenomenon of trirhythmicity in which three stable periodic regimes coexist for the same set of parameter values. The system evolves to either one of these three oscillatory regimes, depending on the initial conditions (Decroly and Goldbeter, 1984). It seems that such phenomenon has not yet been observed in chemical or biological systems.

The possible significance of multiple oscillations is illustrated in Figure 13 which shows how the same perturbation (here, the addition of a sufficient quantity of substrate) can induce the passage from a small-amplitude to a large-amplitude periodic regime, and back, depending on the phase of each oscillation at which the perturbation is applied. This particular example of birhythmicity was recently obtained in a two-variable model (Moran and Goldbeter, 1984) derived from equs. (1) with an additional term describing the recycling of product into substrate in the scheme of Figure 1. In bistability, addition of an intermediate is needed to pass from one stable steady state to the other, for a given set of conditions; removal of the same chemical is then required to achieve the reverse transition. Birhythmicity provides added sensitivity and flexibility to an oscillatory system which could pass from one periodic regime to another, and back, upon reception of the same stimulus, depending on the particular phase at which the system is perturbed.

Complex oscillatory phenomena in the model for the cAMP signalling system based on receptor modification

The complex oscillatory phenomena described above were shown to arise from the interplay between two instability-generating mechanisms, in system comprising two coupled autocatalytic reactions. It was therefore with some surprise that we observed similar pheno-

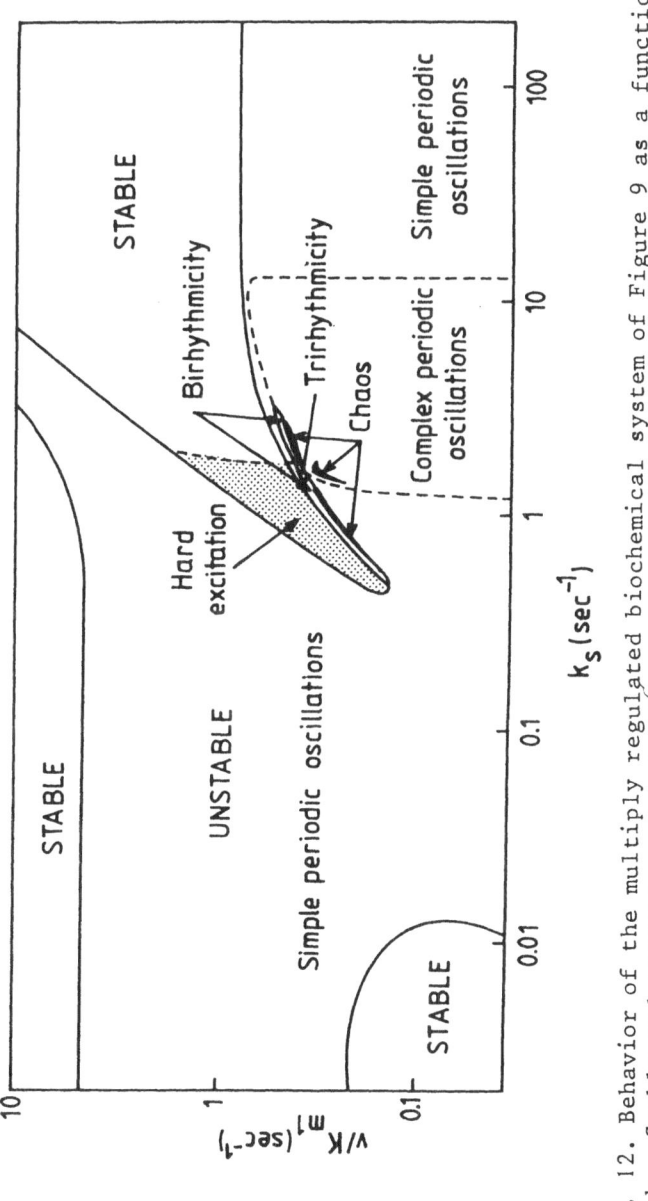

Fig. 12. Behavior of the multiply regulated biochemical system of Figure 9 as a function of v and k_s. Stable and unstable refer to the steady state admitted by equs. (4). The diagram is established for the parameter values of Figure 11a-c.

mena in a model containing a single positive feedback loop.

The model for the cAMP signalling system based on receptor
modification (Figure 3b) is governed by equs. (2). We have shown
that when the steady state admitted by these equations becomes unstable, the system generally undergoes simple periodic oscillations
around the unstable steady state (Figure 6). The first indication
of complex oscillatory behavior was the observation of multiple
peaks of cAMP over a period, giving rise to a phenomenon of bursting
(Figure 14) analogous to that shown in Figure 11d. Aperiodic oscillations occur for closely related values of the parameters. This chaotic behavior is characterized by the evolution towards a strange
attractor in the phase space (Figure 15). Finally, the model also
exhibits birhythmicity. Of the two stable periodic regimes which
coexist for the same set of parameter values, one is a regime of
complex periodic oscillations which corresponds to a folded limit
cycle.

As to the relevance of these results to the development of
D. discoideum, there are no available data that suggest the occurrence of either complex periodic oscillations in the form of bursting,
or birhythmicity. There exists, however, an interesting observation
by Durston (1974) which suggests that chaotic behavior may occur
in a mutant of D. discoideum . By measuring the time intervals separating successive waves of amoebae aggregating on agar, Durston
confirmed that aggregation in the wild type of D. discoideum proceeds
in a periodic manner owing to the periodic, autonomous signalling by
aggregation centers. In the mutant FR17, however, no regular periodicity was found for the successive waves of chemotactically responding amoebae. Centers in this mutant are thus characterized by
"aperiodic autonomous cAMP signalling" (Durston, 1974). Durston
pointedly suggested that "the mutant FR17 may prove to be useful
in investigating D. discoideum autonomous signalling and the basis
of signal periodicity".

A first step in this direction was taken by Coukell and Chan
(1980) who reported about a study in cell suspensions of a tempera-
ture-sensitive mutant derived from FR17. They noted the early
appearance and "erratic" nature of the oscillations of cAMP in this
mutant. To conclude that such oscillations are truly aperiodic would
require a more prolonged study carried over a large number of
cycles, and a comparison with the corresponding oscillations in wild-
type cells. The preliminary observations of Coukell and Chan nevertheless support the view that FR17 may provide an example of chaotic
behavior at the cellular level. It is of interest that the activity
of adenylate cyclase in this mutant is increased as compared with
wild-type cells. In the model, chaos is accordingly observed for
values of the maximum adenylate cyclase activity larger than those
corresponding to periodic oscillations (compare Figures 6 and 15).

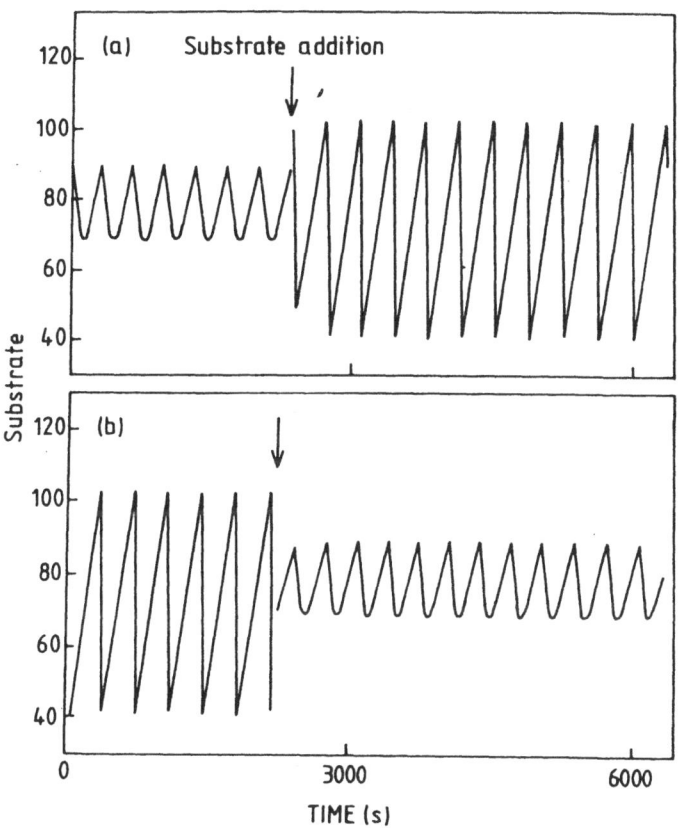

Fig. 13. Birhythmicity. For a given set of parameter values, the two-variable system (Moran and Goldbeter, 1984) derived from Figure 1 with nonlinear recycling of product into substrate, switches between two stable periodic regimes. The switch depends on the phase of oscillations at which the perturbation, in the form of substrate addition, is applied.

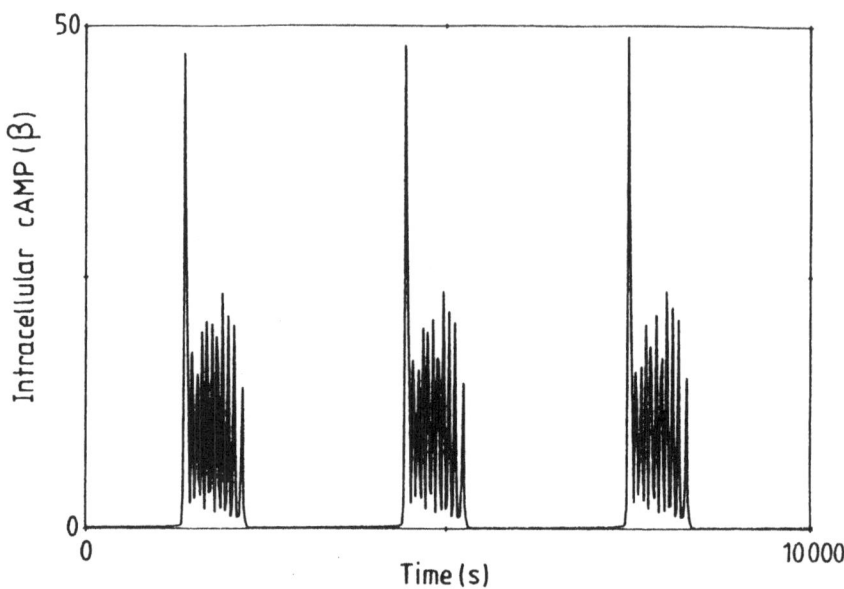

Fig. 14. Bursting in the model for the cAMP signalling system based on receptor desensitization. The time evolution of cAMP is obtained by integration of equs. (3) with $\lambda = \theta = 10^{-2}$; $h=5$; $L_1=398$; $L_2=10^{-1}$; $c=100$; $q=4000$; $\mu=\eta=\varepsilon=10^{-1}$; $\sigma=0.09 \ s^{-1}$; $k_e=0.75 \ s^{-1}$; $k_i=0.2 \ s^{-1}$; $k_t=0.4 \ s^{-1}$; $v=4.7863 \ 10^{-4} \ s^{-1}$; $k_1=10^{-1} \ s^{-1}$; $k_2=2. \ 10^{-2} \ s^{-1}$; $d_1=d_2=d_3= 10 \ s^{-1}$.

The fact that complex oscillatory phenomena occur in the model for the cAMP signalling system seems at variance with the observation that such phenomena originate from the interplay between two instability-generating mechanisms in the model of Figure 9. The contradiction is only apparent. Indeed, two oscillatory mechanisms sharing the same positive feedback loop are hidden in the model for the cAMP signalling system based on receptor modification. The two mechanisms differ by the process responsible for limiting the autocatalysis exerted by cAMP. In the first mechanism, cAMP rises owing to the positive feedback and decreases as a result of substrate limitation when ATP is taken as a variable synthesized at a constant rate. In the second mechanism, the rising phase has the same origin, but the decline in cAMP is brought about by the passage of the receptor in the state uncoupled from adenylate cyclase. Only when these two limiting effects acquire similar importance - which balance can mainly be altered through changes in parameters v, k_1 and k_{-1} in equs. (2) and (3) - do complex oscillatory phenomena appear. Thus bursting, birhythmicity, and chaos occur here as a result of the coupling in parallel of two oscillatory mechanisms, whereas they originate from the coupling in series of two such mechanisms in the model of Figure 9.

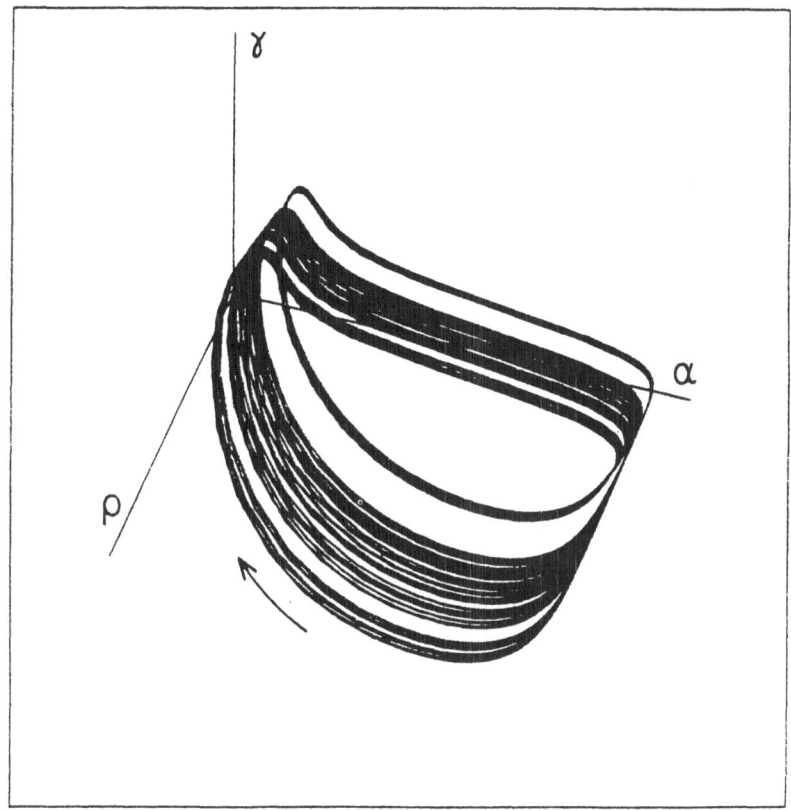

Fig.15. Strange attractor in the chaotic regime admitted by the
seven-variable model (2) based on receptor modification in
D. discoideum. The curve shows the projection in the (α, ρ ,γ)
space of the trajectory obtained by integrating eqs. (3) from zero to
8000s, for $v=0.00074$ s^{-1}. Other parameter values are those of Fig. 6.
The range of variation is 0-1 for ρ,1.81-1.84 for α,and 0-25 for γ.

CONCLUSION

     We have analyzed in the present paper the patterns of ţemporal
self-organization in regulated biochemical systems. Positive feedback is the most likely source of instabilities associated with
self-organization, as exemplified by glycolysis and by cAMP synthesis in D. discoideum . In biochemical and biological systems, periodic behavior is the most common phenomenon encountered beyond the
critical point of instability of the nonequilibrium steady state excitability is a property closely assɔciated with sustained oscillations,
as shown by the analysis of the cAMP signalling system of Dictyostelium amoebae. The models analyzed for this sytem permit to unify
the variety of responses - relay of suprathreshold cAMP pulses,

oscillations, adaptation to constant stimuli - which are observed
in different experimental conditions.

More complex oscillatory phenomena arise from the coupling of
two instability mechanisms within a given system. Thus complex
periodic behavior sometimes accompanied by bursting, aperiodic oscillations (chaos), and multiple stable periodic regimes coexisting for
the same set of parameter values (birhythmicity) were obtained in a
model of two coupled glycolytic oscillators. The coupling in series
of two autocatalytic reactions may at first seem artificial. However,
it could also be taken as a model of an oscillatory system driven by
a periodic force, e.g. a periodic input of substrate, or a periodic
hormonal signal. In physiological conditions, hormones are often
delivered periodically, so that in many instances cells or tissues
endowed with oscillatory properties could be subjected to a periodically changing environment. Experiments and computer simulations
have shown (Boiteux et al , 1975) that the periodic driving of the
glycolytic oscillator could lead to entrainment as well as to complex oscillations.

Neither periodic driving nor multiple positive feedback processes are a prerequisite for complex oscillatory behavior. Our
analysis of a model for the cAMP signalling system based on receptor
modification indicated that such oscillatory phenomena can occur
in this model, although it contains a single positive feedback
loop. Here, new dynamical properties are introduced by the transitions between several receptor conformations. As a result, the
system oscillates along two reaction paths whose interference causes
bursting, chaos, or birhythmicity. The fact that these results were
obtained in a realistic biochemical model based on experimentally
observed properties suggests that complex oscillatory phenomena
may after all be common in metabolic pathways, and in cellular interactions mediated by hormones or neurotransmitters. Bursting
phenomena are observed for the two latter systems. At the biochemical level, complex periodic or aperiodic oscillations have been
observed in glycolysis (See Figure 5 of Hess and Boiteux, 1968,
and Figure 6 of Pye, 1969),in the cAMP synthesis in a mutant of
D. Discoideum (Coukell and Chan, 1980) as well as in the peroxydase
reaction (Olsen and Degn, 1977).

The importance of understanding the molecular basis of such
phenomena, and in particular the basis of aperiodic oscillations,
stems from that the latter have been associated with pathological
conditions in physiological systems (Glass and Mackey, 1979). The
present analysis based on biochemical models suggests that although
chaotic behavior may be common in the space of systems, it is generally found in a restricted domain in the space of parameters.
This conjecture holds with the regularity that characterizes most
biological rhythms.

ACKNOWLEDGEMENTS

One of us (J.L. Martiel) holds a fellowhip from the European Community (Programme "Génie biomoléculaire"). O. Decroly is a fellow from the Institut pour l'Encouragment de la Recherche Scientifique dans l'Industrie et l'Agriculture (IRSIA).

REFERENCES

Alamgir, M. and Epstein, I.R. (1983) J. Am. Chem. Soc. 105, 2500–2501.
Alcantara, F. and Monk, M. (1974) J. Gen. Microbiol. 85, 321–334.
Berridge, M.J. and Rapp, P.E. (1979) J. Exp. Biol. 81, 217–279.
Boiteux, A., Goldbeter, A. and Hess, B. (1975) Proc. Natl. Acad. Sci. USA 72, 3829–3833.
Bonner, J.T. (1967) "The Cellular Slime Molds". Princeton Univ. Press : New Jersey.
Coukell, B. (1981) Differentiation 20, 29–35.
Coukell, M.B. and Chan, F.K. (1980) FEBS Lett. 110, 39–42.
Darmon, M., Barra, J. and Brachet, P. (1978) J. Cell. Sci. 31, 233–243.
Darmon, M. and Brachet, P. (1978) in "Taxis and Behavior", Receptors and Recognition, Ser. B, vol. 5 (Ed. G. Hazelbauer) pp. 103–139. Chapman and Hall : London.
Decroly, O. and Goldbeter, A. (1982) Proc. Natl. Acad. Sci. USA 79, 6917–6921.
Decroly, O. and Goldbeter, A. (1984) C.R. Acad. Sci. Paris Ser. II, in press.
Degn, H. (1968) Nature 217, 1047–1050.
DeHaan, R.L. (1980) in "Current Topics in Developmental Biology", Vol. 16 (Ed. R.K. Hunt) pp. 117–164. Academic Press : New York.
Demongeot, J. and Kellershohn, N. (1983) Lecture Notes in Biomathematics 49, 17–31.
Devreotes, P.N. and Steck, T.L. (1979) J. Cell Biol. 80, 300–309.
Dinauer, M., MacKay, S. and Devreotes, P. (1980) J. Cell Biol. 86, 537–544.
Durston, A.J. (1973) J. Theor. Biol. 42, 483–504.
Durston, A.J. (1974) Develop. Biol. 38, 308–319.
Eschrich, K., Schellenberger, W. and Hofmann, E. (1980) Arch. Biochem. Biophys. 205, 114–121.
Europe-Finner, G.N., McClue, S.J. and Newell, P.C. (1984) FEMS Microbiol. Lett. 21, 21–25.
Feigenbaum, M.J. (1978) J. Stat. Phys. 19, 25–52.
Fessard, A. (1936) "Propriétés Rythmiques de la Matière Vivante", Actualités Scientifiques et Industrielles 417. Hermann : Paris.
Field, R.J., Körös, E. and Noyes, R.M. (1972) J. Am. Chem. Soc. 94, 8649–8664.
Fitzhugh, R. (1961) Biophys. J. 1, 445–466.
Frenkel, R. (1968) Arch. Biochem. Biophys. 125, 151–156.

Geller, J. and Brenner, M. (1978) Biochem. Biophys. Res. Commun. 81, 814–821.

Gerisch, G. (1963) "Entwicklung von Dictyostelium", Film C876T. Institut für den Wissenschaftlichen Film: Göttingen.

Gerisch, G. (1968) Curr. Top. Devel. Biol. 3, 157–197.

Gerisch, G. (1982) Annu. Rev. Physiol. 44, 535–552.

Gerisch, G. and Hess, B. (1974) Proc. Natl. Acad. Sci. USA 71, 2118–2122.

Gerisch, G., Malchow, D., Roos, W. and Wick, U. (1979) J. Exp. Biol. 81, 33–47.

Gerisch, G. and Wick, U. (1975) Biochem. Biophys. Res. Commun. 65, 364–370.

Ghosh, A. and Chance, B. (1964) Biochem. Biophys. Res. Commun. 16, 174–181.

Gingle, A.R. and Robertson, A. (1976) J. Cell Sci. 20, 21–27.

Glass, L. and Mackey, M.C. (1979) Ann. N.Y. Acad. Sci. 316, 214–235.

Glazer, P.M. and Newell, P.C. (1981) J. Gen. Microbiol. 125, 221–232.

Goldbeter, A. (1975) Nature 253, 540–542.

Goldbeter, A. (1980) in "Mathematical Models in Molecular and Cellular Biology" (Ed. L.A. Segel) pp. 248–291. Cambridge Univ. Press.

Goldbeter, A. and Caplan, S.R. (1976) Annu. Rev. Biophys. Bioeng. 5, 449–476.

Goldbeter, A. and Decroly, O. (1983) Am. J. Physiol. (Regul. Integr. Comp. Physiol. 14), 245, R478–R483.

Goldbeter, A. and Erneux, T. (1978) C.R. Acad. Sci. Paris 286 C, 63–66.

Goldbeter, A., Erneux, T. and Segel, L.A. (1978) FEBS Lett. 89, 237–241.

Goldbeter, A. and Koshland Jr., D.E. (1982) J. Mol. Biol. 161, 395–416.

Goldbeter, A. and Lefever, R. (1972) Biophys. J. 12, 1302–1315.

Goldbeter, A. and Martiel, J.L. (1980) Fed. Proc. 39, 1804.

Goldbeter, A. and Martiel, J.L. (1983) Lecture Notes in Biomathematics 49, 173–188.

Goldbeter, A. and Nicolis, G. (1976) in "Progress in Theoretical Biology" (Eds. F. Snell and R. Rosen), vol. 4, pp. 65–160, Academic Press : New York.

Goldbeter, A. and Segel, L.A. (1977) Proc. Natl. Acad. Sci. USA 74, 1543–1547.

Goldbeter, A. and Segel, L.A. (1980) Differentiation 17, 127–135.

Goldbeter, A. and Venieratos, D. (1980) J. Mol. Biol. 138, 137–144.

Van Haastert, P.J.M. and Van der Heijden, P.R. (1983) J. Cell Biol. 96, 347–353.

Heidmann, T. and Changeux, J.P. (1978) Annu. Rev. Biochem. 47, 317–357.

Hers, H.G. and van Schaftingen, E. (1982) Biochem. J. 206, 1–12.

Hess, B. and Boiteux, A. (1968) in "Regulatory Functions of Biological Membranes" (Ed. J. Järnefelt) pp. 148–162. Elsevier : Amsterdam.

Hess, B. and Boiteux, A. (1971) Annu. Rev. Biochem. 40, 237-258.
Hess, B., Boiteux, A. and Krüger, J. (1969) Adv. Enzyme Regul. 7, 149-167.
Higgins, J. (1964) Proc. Natl. Acad. Sci. USA 51, 989-994.
Katz, B. and Thesleff, S. (1957) J. Physiol. 138, 63-80.
Kauffman, S., Shymko, R.M. and Trabert, K. (1978) Science 199, 259-266.
De Kepper, P. (1976) C.R. Acad. Sci. Paris 283 C, 25-28.
Klein, C. (1976) FEBS Lett. 68, 125-128.
Klein, C. (1979) J. Biol. Chem. 254, 12573-12578.
Klein, C. and Darmon, M. (1977) Nature 268, 76-78.
von Klitzing, L. and Betz, A. (1970) Arch. Mikrobiol. 71, 220-222.
Konijn, T.M. (1972) Adv. Cyclic Nucleot. Res. 1, 17-31.
Konijn, T.M., Van de Meene, J.G.C., Bonner, J.T. and Barkley, D.S. (1969) Proc. Natl. Acad. Sci. USA 58, 1152-1154.
Koshland, D.E. Jr. (1979) Physiol. Rev. 59, 811-862.
Koshland, D.E. Jr., Goldbeter, A. and Stock, J.B. (1982) Science 217, 220-225.
Loomis, W.F., ed. (1982) "The Development of Dictyostelium discoideum". Academic Press : New York.
Lubs-Haukeness, J. and Klein, C. (1982) J. Biol. Chem. 257, 12204-12208.
Malchow, D., Böhme, R. and Gras, U. (1982) Biophys. Struct. Mech. 9, 131-136.
Monod, J., Wyman, J. and Changeux, J.P. (1965) J. Mol. Biol. 12, 88-118.
Moran, F. and Goldbeter, A. (1984) Biophys. Chem., in press.
Naparstek, A., Romette, J.L., Kernevez, J.P. and Thomas, D. (1974) Nature 249, 490-491.
Newell, P.C. and Ross, F.M. (1982) J. Gen. Microbiol. 128, 2715-2724.
Nicolis, G. and Prigogine, I. (1977) "Self-Organization in Nonequilibrium Systems". Wiley : New-York.
Olsen, L.F. and Degn, H. (1977) Nature 267, 177-178.
Pacault, A., Hanusse, P., De Kepper, P., Vidal, C. and Boissonade, J. (1976) Acc. Chem. Res. 9, 438-445.
Prigogine, I. (1967) "Introduction to Thermodynamics of Irreversible Processes". Wiley : New-York.
Pye, E.K. (1969) Can. J. Bot. 47, 271-285.
Rapp, P.E. and Berridge, M.J. (1977) J. Theor. Biol. 66, 497-525.
Robertson, A. and Drage, D.J. (1975) Biophys. J. 15, 765-775.
Roos, W., Nanjundiah, V., Malchow, D. and Gerisch, G. (1975) FEBS Lett. 53, 139-142.
Roos, W., Scheidegger, C. and Gerisch, G. (1977) Nature 266, 259-261.
Schulmeister, T. and Sel'kov, E.E. (1978) Stud. Biophys. 72, 111-112, and Microfiche 1/24-37.
Sel'kov, E.E. (1968) Eur. J. Biochem. 4, 79-86.
Shaffer, B.M. (1962) Adv. Morphogen. 2, 109-182.

Smith, W.R. (1983) Am. J. Physiol. (Regul. Integr. Comp. Physiol. 14) 245, R473-R477.
Springer, M.S., Goy, M.F. and Adler, J. (1979) Nature 280, 279-284.
Stadel, J.M., Nambi, P., Shorr, R.G.L., Sawyer, D.F., Caron, M.G. and Lefkowitz, R.J. (1983) Proc. Natl. Acad. Sci. USA 80, 3173-3177.
Theibert, A. and Devreotes, P.N. (1983) J. Cell Biol. 97, 173-177.
Tolkovsky, A.M., Braun, S. and Levitki, A. (1982) Proc. Natl. Acad. Sci. USA 79, 213-217.
Tornheim, K. and Lowenstein, J.M. (1974) J. Biol. Chem. 249, 3241-3247.
Venieratos, D. and Goldbeter, A. (1979) Biochimie 61, 1247-1256.
Vidal, C. and Pacault, A., Eds. (1981) "Nonlinear Phenomena in Chemical Dynamics". Series in Synergetics, Vol. 12, Springer : Berlin.
Winfree, A.T. (1972) Science 175, 634-636.
Winfree, A.T. (1974) Sci. Amer. 230, 82-95.
Winfree, A.T. (1980) "The Geometry of Biological Time". Springer : New-York.

DYNAMIC COUPLING AND TIME-PATTERNS OF GLYCOLYSIS

Benno Hess, Dietrich Kuschmitz and Mario Markus

Max-Planck Institut für Ernährungsphysiologie
Dortmund, West-Germany

DYNAMIC ORGANIZATION

The property of self-organization is a fundamental feature of living systems. Macroscopically, it is reflected in the phenomena of evolution, of differentiation and of numerous other biological functions. The quality of organization results from basic thermodynamic and kinetic constraints, to which the occurrence of biological systems is fundamentally bound, and its theoretical frame lies in the concept of dissipative structure as a new science of motion.

Biological order results from the initiation of chemical transformations and transport phenomena that are maintained far from equilibrium; these processes are open by a continuous supply of energy and matter and quasi-closed with respect to a large number of chemical structures (such as lipids, proteins, and nucleic acids) in a unicellular or multicellular scenario. The "far-from-equilibrium conditions" implies the irreversibility of the processes and unidirec-

---

ABBREVIATIONS

PK, pyruvate kinase (EC 2.7.1.40); PFK, phosphofructokinase (EC 2.7.1.11); FBP, fructose-1,6-bisphosphate; F6P, fructose-6-phosphate; PEP, phosphoenolpyruvate; $[K_{tot}]$ and $[Mg_{tot}]$, total concentrations (bound plus unbound) of potassium and magnesium

tionality of the evolution of time. Under equilibrium conditions
all living phenomena disappear. Biological systems consist of
complex chemical networks of interactions of a vast number of different chemicals, resulting in highly nonlinear kinetics. The kinetic
nonlinearity arises from different sources :

1 . Multiple types of positive or negative feedback interactions
are organized in the form of enzymatic cycles such as energy metabolism, biosynthetic pathways, or the predicted evolutionary hyper-
cycles.

2. Many processes are controlled by allosteric enzymes responding to
small changes in substrate, products, and controlling ligands in a
cooperative manner by changing their conformation states.

3. The organization of transport of chemical particles in living
systems implies not only free diffusion but to a large extent processes
that occur upon coupling between vectorial transmembrane events and
enzyme functions. The activity of membrane-bound enzymes is controlled not only by the components of the membrane itself but, in addition,
by the nature of the membrane protein-directed diffusion processes.
Electrical fields have been observed to be an additional large-
scale force affecting the transport of charged particles and controlling the microenvironment of catalytic centers.

Thermodynamic conditions and nonlinear chemical transformation and transport processes result in the evolution of a number of
dynamic states such as :

1. multiple steady states with monotonic or overshooting transitions from one state to another (stable node or stable focus, respectively),

2. rotation of a limit cycle around an unstable singular point
(several limit cycles are possible),

3. chaotic states.

In the context of this presentation we would like to discuss
the problem of dynamic coupling between regulated biochemical
processes. The intricate relationships in cellular processes yield
complex dynamic behaviour, which can only be understood by reducing
the objects of study to a very simple level. Drawing primarily on
our own experiments, we would therefore like to dwell in the problem of time pattern formation in glycolysis under periodic excitation, and the coupling of glycolysis to the proton translocation process in the cell membrane. The next section will deal with glycolytic
time patterns on the basis of model analysis, although experimental
data have been obtained recently.

GENERATION AND CONTROL OF TIME PATTERN IN A GLYCOLYTIC MODEL

In earlier investigations, the overall dynamics of non-oscil-latory as well as oscillatory glycolysis have been studied experimentally in yeast cells as well as cell free extracts of yeast. Concomitant studies of mathematical models of glycolysis led to the recognition, that the dynamic behaviour of this process can well be reduced to the kinetic properties of the allosteric enzyme phosphofructokinase, which functions as the master oscillator of the whole process, and the enzyme pyruvate kinase, catalyzing the sink reaction (Boiteux et al, 1975; Plesser, 1977). Experimentally, phase sensitivity, phase shifting properties, control of amplitude and frequency have been observed. For oscillating conditions, a limit cycle behaviour of glycolysis was found well in the physiological turnover range. Of special interest is the ability of synchronization of glycolytic oscillations with an external periodic substrate addition. Entrainment by the fundamental, the 1/2 harmonic and the 1/3 harmonic has been observed. In similar experiments, the response of glycolytic oscillations toward stochastic perturbations has also been described (Boiteux et al, 1975).

In order to determine the extent and quality of the domains of entrained and non-entrained oscillations of glycolysis subject to external forcing, we analyzed numerically a system consisting of the two enzymes phosphofructokinase (PFK, EC 2.7.1.11) and pyruvate kinase (PK, EC 2.7.1.40), where fructose-6-phosphate (F6P) and phosphoenolpyruvate (PEP) are supplied as periodic input. For preliminary simplification, the effect of a feedback process of glycolysis on the substrate input (see next section) is neglected. The system is described by the following equations :

(1)
$$\frac{d[F6P]}{dt} = \bar{V}_{in} + A \sin \omega_e t - V_{PFK}$$

(2)
$$\frac{d[PEP]}{dt} = \bar{V}_{in} + A \sin \omega_e t - V_{PK}.$$

(3)
$$\frac{d[ADP]}{dt} = V_{PFK} - V_{PK}$$

(4)
$$\frac{d[ATP]}{dt} = V_{PK} - V_{PFK}$$

The input flux is given by $\bar{V}_{in}$ + A sin ω_e t, where $\bar{V}_{in}$ is a constant, A is the amplitude of the external excitation and ω_e its frequency. The products fructose-1,6-bisphosphate (FBP) and pyruvate (Pyr) ac-cumulate in a concentration range not effecting the enzyme kinetics i.e. product inhibition can be ignored. We set $[FBP] = 0$ at t = 0. V_{PFK} and V_{PK} are the enzymic rates. Detailed rate laws have been derived from experiments with PFK (Blangy et al, 1968) and PK (Boiteux et al, 1983) from E. coli for the relevant concentration space of enzymic ligands. The rate law for PK obeys a two-substrate kinetic mechanism. The rate law for PFK was generalized to a two-substrate mechanism by assuming random substrate binding with dis-sociation constants independent of the binding order. Interference of the enzyme by fructose-2,6-bisphosphate was not considered. The rate laws include the activation of PFK by ADP, the inhibition of PFK by PEP, and the inhibition of PK by ATP. In addition, the effect of Mg^{2+} on the PK kinetics and of complex formation of cations with ligands were considered as described by Markus et al (1980).

Equations (1) to (4) lead to the two conservation laws $[PEP] + [ATP] - [F6P] = C$ and $[ADP] + [ATP] = ADN$, where C and ADN are constant bifurcation parameters of the system. The dynamics of the system depends on the following additional parameters :
$\nu = \bar{V}_{in}/V_{max(PK)}$, $\alpha = A/V_{max(PK')}$, $\varepsilon = V_{max(PFK)}/V_{max(PK)}$, ω_e, $[Mg_{tot}]$ and $[K_{tot}]$ (for details see Hess and Markus, 1983).

At constant input rate (A = 0), only two kinds of attractors are found : steady states and limit cycles. At a periodic input rate (A, $\omega_e \neq 0$), other attractors can be reached : periodic orbits due to entrainment as examined in Boiteux et al (1975), quasiperio-dic oscillations and chaos. The solutions can be displayed in a spa-ce where two dimensions are metabolite concentrations and the third dimension is time. Such a display leads to a trajectory screwing its way into infinity in the direction of the time axis, as exemplified for a chaotic solution in Fig. 1. A different form of three-dimensional display permitting the solution to be shown in a closed form is given in Fig. 2. In this representation , we plot two metabolite concentrations ($[ADP]$ and $[PEP]$ in this case) on a plane and let this plane rotate around the $[PEP]$-axis so that one rotation is completed after one input flux period T_e. Another dis-play method, revealing the orderliness of chaotic dynamics in a two dimensional representation, is the stroboscopic phase portrait, where the metabolite concentrations are plotted at equidistant values of time. Fig. 3a shows such a portrait (dotted "islands"), the distance between plotted points being T_e. This portrait corresponds to the intersection of the attractor shown in Fig. 2 with the $[PEP] - [ADP]$ cos $\omega_e t$ -plane.

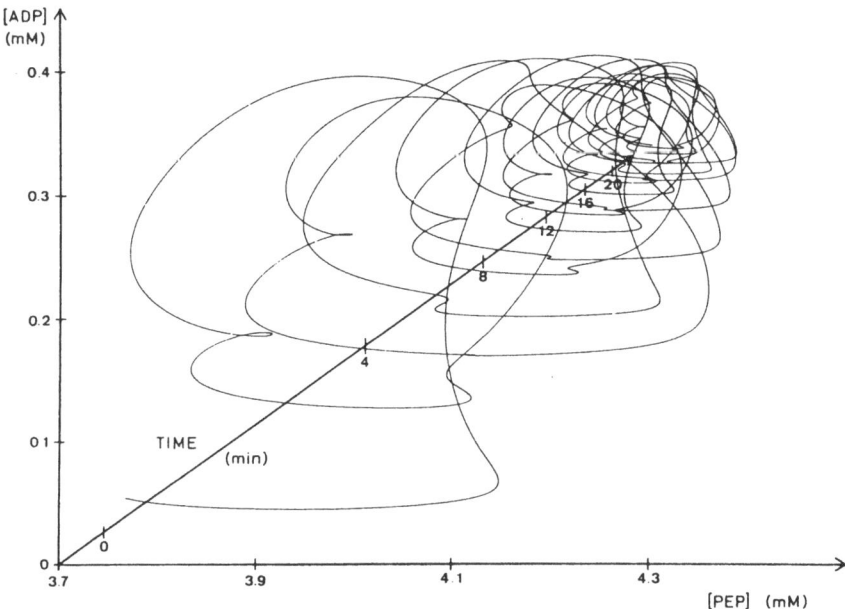

Fig.1. Chaotic orbit. The conditions are : ν =0.12, ε= 0.34, ADN = 2.55 mM, C = 5.75 mM, $\left[Mg_{tot}\right]$ = 5 mM, $\left[K_{tot}\right]$ = 200 mM , A = 1.14mM/min, ω_e = 14 min^{-1} (time dimensioning was set assuming $V_{max(PFK)}$=3.3 mM/min).

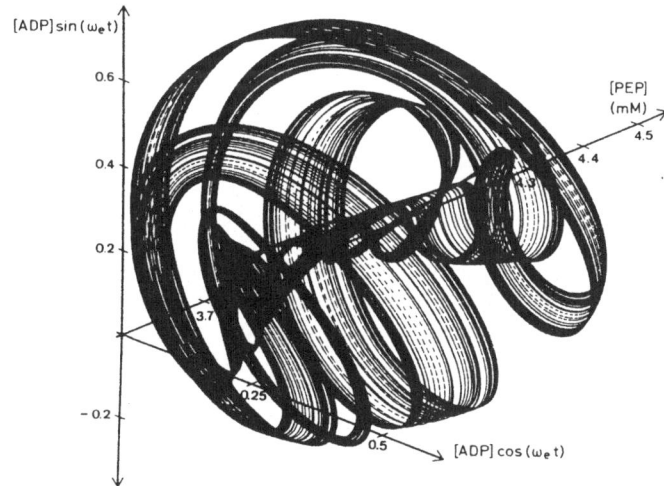

Fig. 2. Closed representation of the chaotic orbit shown in Fig.1. All scales are linear.

Due to the two conservation laws given above, the four metabolite concentrations $[ADP]$, $[ATP]$, $[PEP]$ and $[F6P]$ can be displayed two-dimensionally along the edges of a trapezium, as shown in Figs. 3a and 3b. The trapezium is constructed as an equilateral triangle truncated by a line parrallel to one of its sides. The metabolite concentrations can be read at each point of the trajectory as shown by the dashed lines in Figs. 3a and 3b. This representation is an adaptation to our   problem of the known method of simplex portraits (Hess and Markus, 1983; Schuster et al, 1979). In Fig. 3a we also show an attractor with period $12T_e$ (continuous line). The two attractors shown in Fig. 3a coexist in phase space under the same values of bifurcation parameters. Such a phenomenon has been called "Birhythmicity" (Decroly and Goldbeter, 1982).

Solutions can be switched in phase space from one coexisting attractor to another by pulsed metabolite additions (or substractions) under the restriction of constant bifurcation parameters. In such a switching process, transients can be avoided by choosing the right time and the right amounts of added (or substracted) metabolites, so that the system can jump directly from one orbit into the other (Hess and Markus, 1983).

Fig. 3b shows the coexistence of a quasiperiodic oscillation (dots) and a solution with period T_e (continuous line). In order to investigate the variety of dynamic states as a function of the amplitude, we varied A as shown in Fig. 4 using the "slow-time" method (Minorsky , 1974; Kubiček and Marek, 1983) holding the other parameters constant at the values corresponding to Fig.3b. Fig. 4 was obtained by changing A back and forth at a rate $\Delta A/A = \pm 3 \times 10^{-5}$ per input flux period. This procedure reveals in Fig. 4 a complex scheme consisting of interwoven hysteresis loops.

Fig. 5 shows the ADP concentrations corresponding to Fig.4. The concentrations are plotted one input flux period apart. A solution with period nT_e implies n ADP concentrations for each value of A and thus leads to a set of n lines in the figure (for example : period $2T_e$ for A > 0.375 mM/min or a period $3T_e$ for A < 0.391 mM/min). Spread out points in the figure imply a quasi-periodic solution (for example : A = 0.349 or 0.356 mM/min.). Oscillating curves imply either a quasiperiodic or a periodic solution. Further diagnostic is needed in the last case : the solution is periodic if a finite number of distinct curves remain when the value of $\Delta A/A$ per input flux period tends to zero (for example period $5T_e$ at 0.351< A < 0.354 mM/min) and quasiperiodic otherwise (for example at 0.365 mM/min). Transitions are characterized by narrow zones of aperiodic behaviour, as for example around A = 0.391 mM/min. In the interval 0.357 mM/min< A < 0.36 mM/min, Figs. 4 and 5 reveal the coexistence of four attractors in phase space

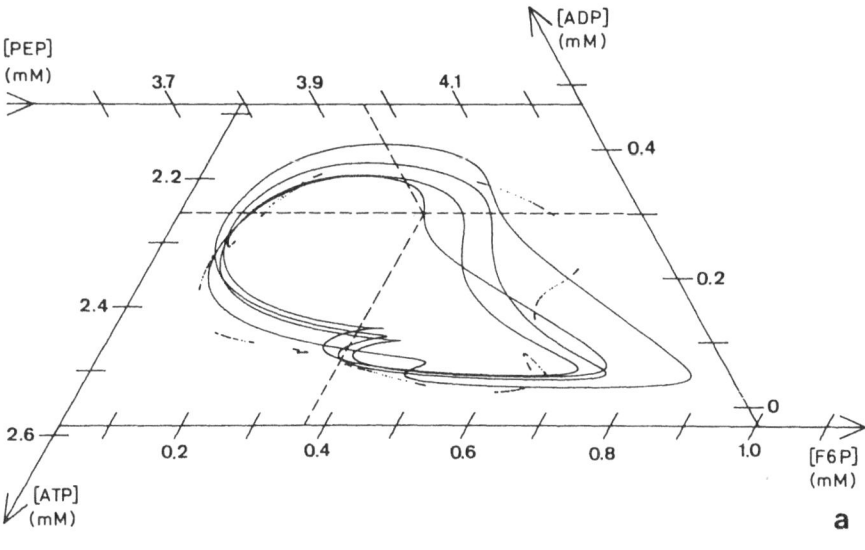

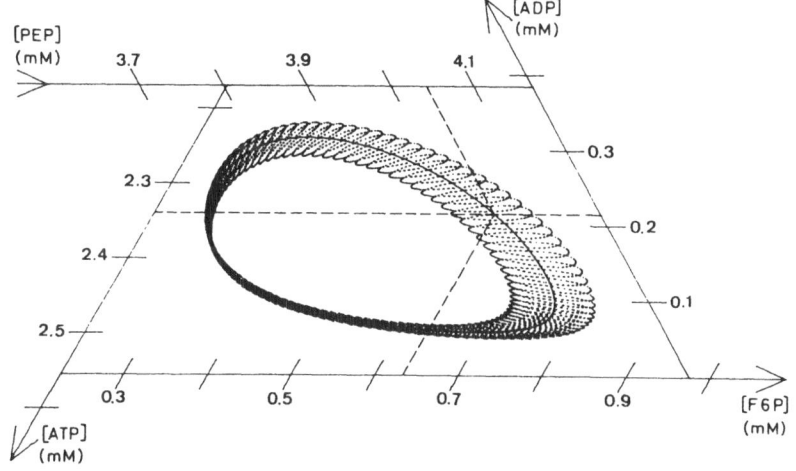

Fig. 3. a) Coexistence of an attractor with period 12 times the input flux period (continuous line) with the chaotic attractor shown
in Figs. 1 and 2 (dots). While the continuous line contains all the
points of the attractor, the dots are plotted in a stroboscopic representation at times one input flux period apart.
b) Coexistence of an attractor with period equal to the input flux
period (continuous line) with a quasiperiodic oscillation (dots).
The conditions are those of Fig. 1, except for A = 0.36 mM/min and
$\omega_e = 6$ min^{-1} . The dots are plotted in a stroboscopic representation
each hundredth of the input flux period.

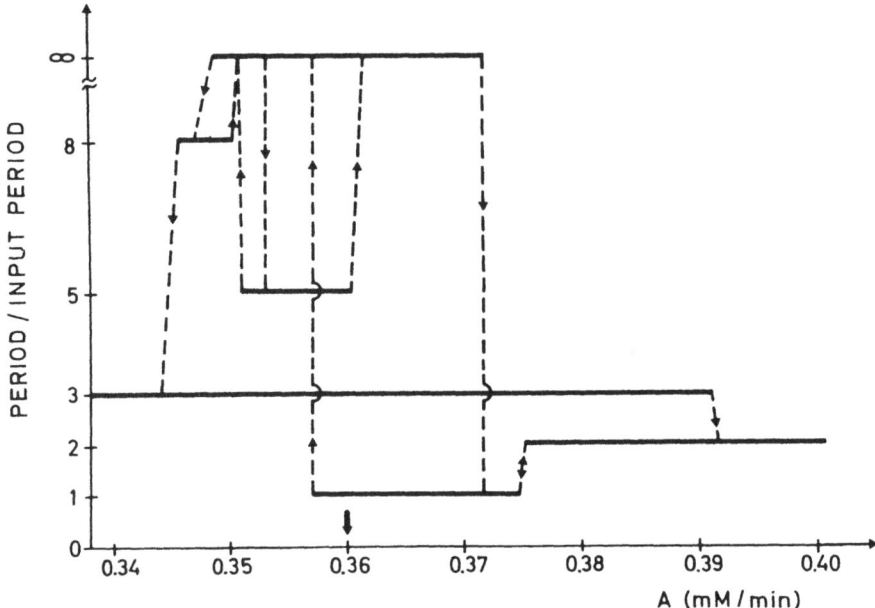

Fig.4. Complex hysteretic response of the system in dependence of
the input flux amplitude. The qualitative response is indicated
in the ordinate by the oscillation period divided by the input
flux period. Period ∞ indicates quasiperiodicity. The transitions
can only take place in the directions given by the arrows. The
conditions are those of Fig. 3b except that A is varied . The
arrow pointing to the abscissa indicates the amplitude corresponding to Fig. 3b.

("tetrarhythmicity"), their periods being T_e, $3T_e$, $5T_e$ and ∞
(quasi-periodicity). Further examples of conditions with tetrarhytmic
performance are : periods $3T_e$, $10T_e$, $12T_e$ and $20T_e$ at A = 1.11 mM/
min and $\omega_e = 14$ min^{-1}; periods $2T_e$, $5T_e$, $7T_e$ and ∞ (chaos) at
A = 0.437 mM/min and $\omega_e = 6.5$ min^{-1} (the other conditions are given
in the caption of Fig. 1).

        A scan of the space of bifurcation parameters reveals a rich
variety of periodic, quasiperiodic and chaotic solutions. When the
metabolite concentrations corresponding to these solutions are
plotted against time, patterns differing in amplitude and (predominantly) in its modulation are obtained (Hess and Markus, 1983). We
may call the variety of time patterns that a system might display
the "pattern library" of the system. A single pattern in such a

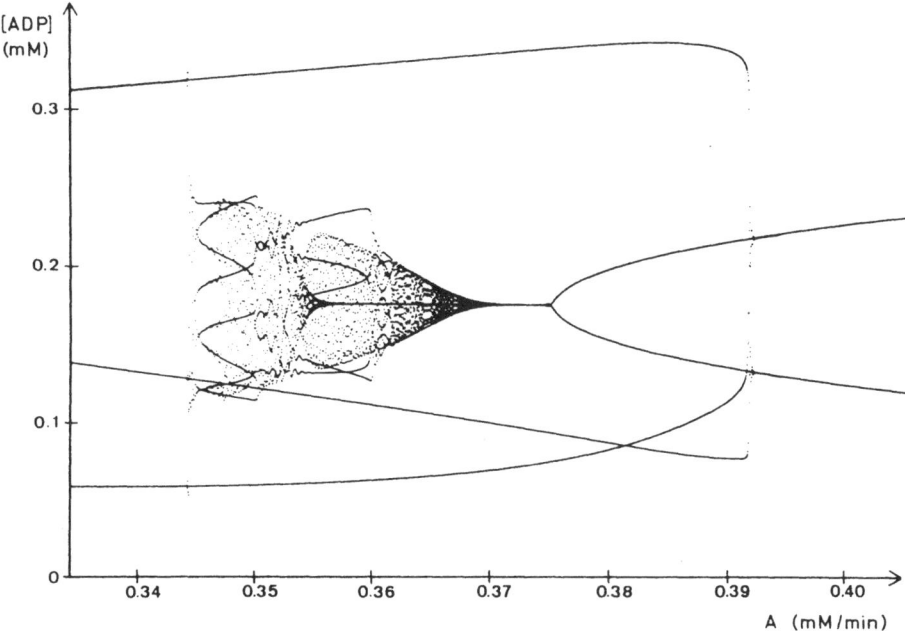

Fig.5. ADP concentrations corresponding to Fig. 4. The values of
the concentrations are plotted one input flux period apart.

a library is accessible in two ways : 1. by varying a bifurcation
parameter, as exemplified in Fig.4 when variations are performed
in horizontal direction, i.e. by varying the abscissa value and
2. by changing the metabolite concentrations through additions or
substractions at constant bifurcation parameters, so that a jump
is caused between attractors coexisting in phase space. This
second possibility can be understood in the context of Fig. 4 by
jumping in vertical direction from one horizontal line to the
other. These model experiments illustrate the potency for time
pattern generation and control that is observed in an allosteric
enzyme model coupled chemically to another frequency generator.

   Recently non-periodic oscillations have been measured by
NADH fluorescence (ϕ) in extracts of S. cerevisiae in a range of
input flux amplitudes and frequencies, which was predicted by this
model analysis. Evidence that these oscillations are chaotic was
obtained by : 1). Power spectra analysis, showing higher broadband
noise than entrained oscillations, 2). Stroboscopic plots in the
$\phi - d\phi/dt$-plane revealing typical qualities of strange attractors,
3). Stroboscopic transition functions $\phi(n+1)$ versus $\phi(n)$,(n= 1,2...)
admitting period three and thus implying chaos according to the
Li-Yorke theorem (Li and Yorke, 1975).

SYNCHRONIZATION OF OSCILLATING GLYCOLYSIS AND OSCILLATING MEMBRANE
POTENTIAL IN INTACT YEAST

In order to gain insight into the modes of dynamic coupling
we analyzed the interaction of glycolysis and the cellular plasma
membrane potential as demonstrated schematically in Fig. 6 (Hess et
al, 1983). In case of suppressed respiration through the absence
of oxygen and in the presence of glucose, glycolysis is the only
ATP-generating system and drives the proton translocating ATPase
of the plasma membrane (Goffeau and Slayman, 1981), which builds
up a proton gradient and, with or without companion movement of
other ions, sets up a plasma membrane potential. It is interesting
to note that the membrane potential generated might well affect the
glucose uptake in an additional control loop combining scalar and
vectorial enzymes (Hess et al, 1983).

The simultaneous time analysis of the change of the activity
of glycolysis and the plasma membrane potential is based on a
record of NADH fluorescence as indicator of glycolysis and rhodamin-
6G fluorescence (Aiuchi et al, 1980; Aiuchi et al, 1982) as indicator of plasma membrane potential. From independent experiments it
is known that fluorescence quenching indicates increase in the
plasma membrane potential (Aiuchi et al, 1980; Aiuchi et al, 1982).
In order to exlude interference of respiration, the experiments
described here are carried out with respiratory deficient mutants
of yeast. For recording, the double fluorimetry technique was used
(Hess et al, 1983).

Upon activation of a suspension of yeast cells by addition
of glucose first a short transient is observed in the record of
NADH fluorescence as well as rhodamin-6G fluorescence . Then glycolysis reaches an oscillatory state somewhat before an oscillating
state of the plasma membrane potential shows up. In a typical record,
steady oscillations of glycolysis as well as of plasma membrane is
observed (see Fig. 7a). The membrane potential oscillates between
50 and 100% of total rhodamin-6G fluorescence and the NADH-fluores-
cence as indicator of glycolysis oscillates between 80 and 100%
with a period of approximately 60 secs for both components. Comparing
the maxima of the two oscillations it is obvious that both components
are running with the same frequency, and with approximately the
same phase (Hess et al, 1983).

The relationship of NADH and rhodamin fluorescence over one
period is given in the phase plane plot of Fig. 7b. Here we can
distinguish three different phases (I, II, III), each defining
different dynamic states during a limit cycle. Phase II (corresponding to the inital part of the transition from NADH minimum to
NADH maximum), is characterized by a rate of NAD reduction wich is
relatively slow compared to the rate of potential decrease. Conver-

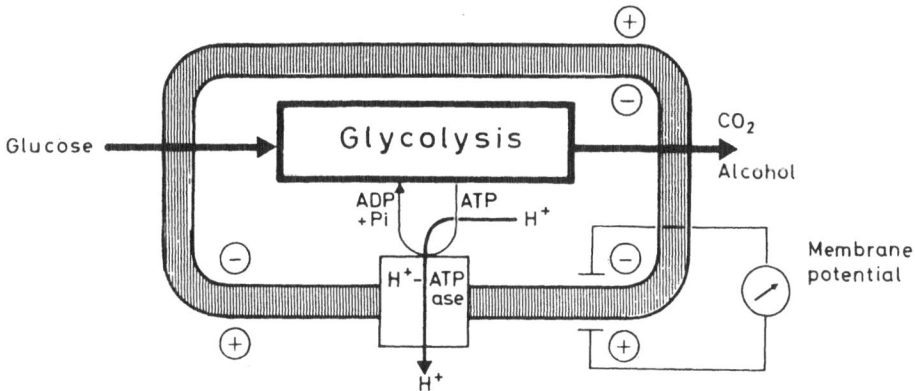

Fig. 6. Scheme of glycolysis and plasmamembrane potential formation
in yeast cells.

sely, in phase III (at the end of the transition towards the NADH
maximum), glycolysis runs fast relative to the potential decrease
rate. In phase I, both processes are synchronized. At the present
time, a complete designation of the rate controlling steps in both
coupled processes is not at hand (Hess et al, 1983).

Evidence that the plasma membrane proton translocating
ATPase is directly involved in the generation of the rhodamin-6G
indicated plasma membrane potential comes from a study in which
the two processes have been uncoupled by proper inhibitors.
Addition of the specific inhibitor of the plasma membrane proton
translocating ATPase, namely diethylstilbestrol, as well as uncoupling with pentachlorophenol, readily inhibits the oscillation
of the plasma membrane potential. On the other hand, the influence
of the proton translocating system on glycolysis can be documented
by affecting the plasma membrane potential from outside the cell
using appropriate cations. Indeed, under non-oscillating conditions
potassium, calcium and lanthanum ions induce oscillations of glycolysis and the plasma membrane potential. A typical experiment in
which lanthanum initiates the oscillatory state of the plasma membrane potential, obviously via interfering with the surface potential, and secondarily drives glycolysis into the oscillatory state
as shown in Fig. 8.

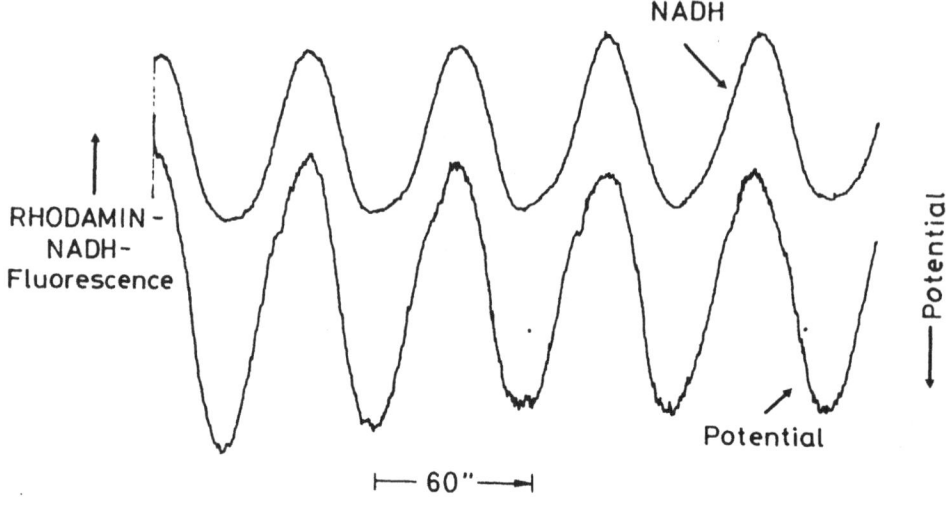

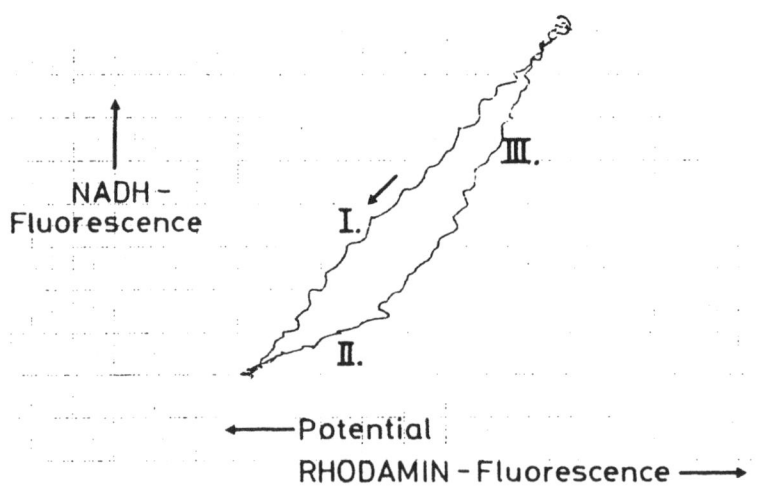

Fig. 7. a) Simultaneous record of oscillations of NADH and of the plasma membrane potential (indicated by rhodamin-6G fluorescence) in yeast cells initiated by the addition of 100 μMoles glucose. b) Phase plane analysis of NADH and rhodamin-6G fluorescence over one period.

It could well be that we are dealing here with a phenomenon of chemical "resonance" between two highly non-linear processes. Indeed, besides the well-known sigmoidal responses in glycolysis, it is known from studies in Neurospora, that the proton translocating plasmamembrane ATPase exhibits a Hill coefficient of 2 (Goffeau and Slayman, 1981) which would be expected for a cooperative allosteric enzyme. The interplay of the two non-linear systems might bring about a plethora.of diverse phenomena of temporal patternization.

The experiments illustrate a novel insight into the dynamic coupling as a time-dependent phenomenon being of interest for understanding a number of other cellular glycolyzing systems for which periodic states are known such as the smooth muscle, the heart muscle as well as neural systems. Future experiments will have to be designed to unveil the mechanism of time pattern control under such in vivo conditions, when a large cellular network is coupled by an intercellular communication system. The studies of the slime mould Dictyostelium discoideum presents already a good example of investigation of such dynamic interactions.

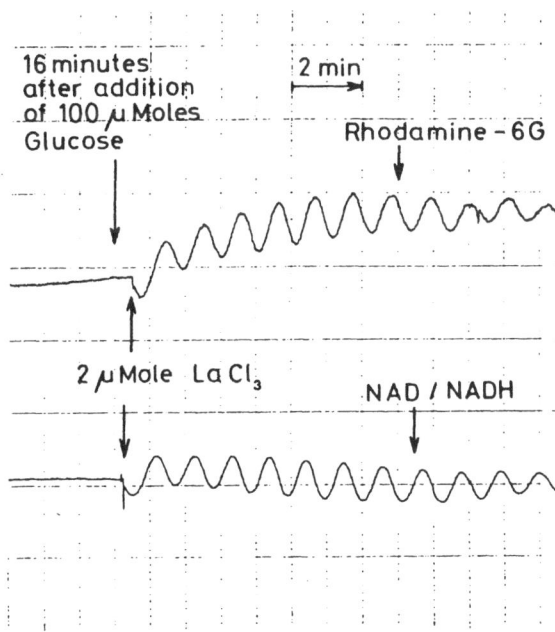

Fig. 8. Initiation of NADH and plasmamembrane potential oscillations by the addition of lanthanum ions.

ACKNOWLEDGEMENTS

    We would like to thank Mrs. Plettenberg for her excellent
typing and editing of the manuscript. Furthermore, we acknowledge
the help in the development of the computer graphics by Mr. H.
Becher and the most generous support of our work by the Fond Chemie,
Frankfurt am Main and the IBM Deutschland GmbH, Stuttgart.

REFERENCES

Boiteux, A., Goldbeter, A. and Hess, B. (1975) Proc. Natl. Acad. Sci.
USA 72 (10), 3829-3833.
Plesser, Th. (1977) in : VII Int. Konf. über nichtlineare Schwingungen
G. Schmidt, ed., Akademie Verlag, Berlin, 2, 273-280.
Blangy, D., Buc, H. and Monod, J. (1968) J. Mol. Biol. 31, 13-35.
Boiteux, A., Markus, M., Plesser, Th., Hess, B. and Malcovati, M.
(1983) Biochem. J. 211, 631-640.
Markus, M., Plesser, Th., Boiteux, A., Hess, B. and Malcovati, M
(1980) Biochem. J. 189, 421-433.
Hess, B. and Markus, M. (1983) in Synergetics : from microscopic to
macroscopic order, Springer Verlag, Berlin, Heidelberg, New York,
pp. 6-16.
Schuster, P., Sigmund, K. and Wolff, R. (1979) SIAM J. Appl. Math.
C 37 (1) 49-54.
Decroly, O. and Goldbeter, A. (1982) Proc. Natl. Acad. Sci. 79,
6917-6921.
Minorsky, N. (1974) Nonlinear oscillations , R.E. Krieger Publ. ,
Huntington, N.Y.
Kubíček, M. and Marek, M. (1983) Computational Methods in bifurcation
theory and dissipative structures , Springer Verlag, New-York, Berlin,
Heidelberg, Tokyo.
Li, T.Y. and Yorke, J.A. (1975) Amer. Math. Mon. 82, 985-992.
Markus, M. and Hess, B. (1984) Proc. Natl. Acad. Sci. USA,
in press.
Hess, B. , Boiteux, A. and Kuschmitz D. (1983) in Biological oxi-
dations, Springer Verlag, Berlin, Heidelberg, New York, pp. 249-266.
Goffeau, A. and Slayman, C.W. (1981) Biochim. Biophys. Acta 639,
197-223.
Aiuchi, T., Tanabe, H., Kurihara, K. and Kobotake, Y. (1980)
Biochim. Biophys. Acta 628, 355-364.
Aiuchi, T., Daimatsu, T., Nakayaka, K and Nakamura, Y. (1982)
Biochim. Biophys. Acta 685, 289-296.

NONEQUILIBRIUM THERMODYNAMICS OF BIOLOGICAL ENERGY CONVERSIONS

J.W. Stucki

Pharmakologisches Institut der Universität
Friedbühlstrasse 49, CH-3018 Bern, Switzerland

INTRODUCTION

Most of the contributions of this symposium are dedicated to a detailed study of the control and regulation of enzymatic mechanisms. Clearly the method of choice to tackle such questions is kinetics. Once all enzymatic parameters such as rate constants or more globally affinities, allosteric constants and the like are known for an enzyme catalyzed reaction, then the dynamics of the system can be completely described. This remarkable success of kinetic analyses for predicting and explaining the dynamics of a complex reaction sequence is in sharp contrast to its notorious inability to deal with energetic questions. This appears perhaps not so astonishing by realizing that kinetics is entirely concerned with flows whereas energetics, in addition, also needs information about the forces involved in the reactions of the system. Here, clearly, a formalism is required which can cope with both quantities, flows as well as forces. The formalism of choice to approach energetic questions is therefore nonequilibrium thermodynamics.

The big advantage of this formalism is that it is purely phenomenological in nature and does thus not require the knowledge of a detailed reaction mechanism. Hence nonequilibrium thermodynamics allows the treatment of the energetics of complex systems on a very general level. On the other hand, nonequilibrium thermodynamics is unable to prove or disprove a proposed reaction mechanism, simply because thermodynamics does not care about mechanisms. Thermodynamics can only tell whether a proposed mechanism violates some basic physical laws such as the second law of thermodynamics. The kineticist, however, is interested in more specific information

because he would like to prove a reaction mechanism. It is probably
for this reason that most kineticists, still, consider nonequilibrium thermodynamics rather as some piece of a nuisance than as
a method providing substancial support to the solution of their
problems.

   The aim of this contribution is twofold : first, I would like
to demonstrate that nonequilibrium thermodynamics can help to
obtain insight into bioenergetic problems which can not be obtained
from a purely kinetic study alone. Second, I would like to show that
judiciously chosen thermodynamic functions are even able to predict
transient kinetics to an astonishing level of detail. This may
appear surprising since according to a standing dogma thermodynamics
is unable to yield information about the velocity of a reaction.
It must be realized, however, that his restriction applies only to
equilibrium thermodynamics which deals with an equilibrium situation,
characterized by the absence of net flows. In contrast, nonequilibrium thermodynamics considers both flows and forces, arbitrarily
far from equilibrium and is therefore entitled to say something about
reaction velocities, not only at a steady state but at a transient
situation as well.

LINEAR ENERGY CONVERTERS, CONDUCTANCE MATCHING, DEGREES OF COUPLING
AND THERMODYNAMIC BUFFERING EXPLAINED WITH OXIDATIVE PHOSPHORYLATION

   Oxidative phosphorylation can be described in a phenomenological manner as a black box to which an input is applied and whose
output is connected to a load conductance as shown in Fig. 1.
The input consists of the oxidizable substrates. It is characterized by the redox potential X_o with an associated net flow oxygen
consumption J_o. The output is, similarly, characterized by an output force, the phosphate potential X_p with an associated net flow
of ATP production J_p. The load conductance summarizes the cellular
ATP utilizing processes. Since we are interested only in a phenomenological description, all we have to verify experimentally is how
the output is related to the input and we can essentially ignore
the internal gearing and wiring of the energy converter. An experimental study with mitochondria isolated from rat liver has
shown that the flows J_p and J_o are linearly related to the output
force X_p according to the relations

(1) $$J_p = L_p X_p + L_{po} X_o$$

(2) $$J_o = L_{po} X_p + L_o X_o$$

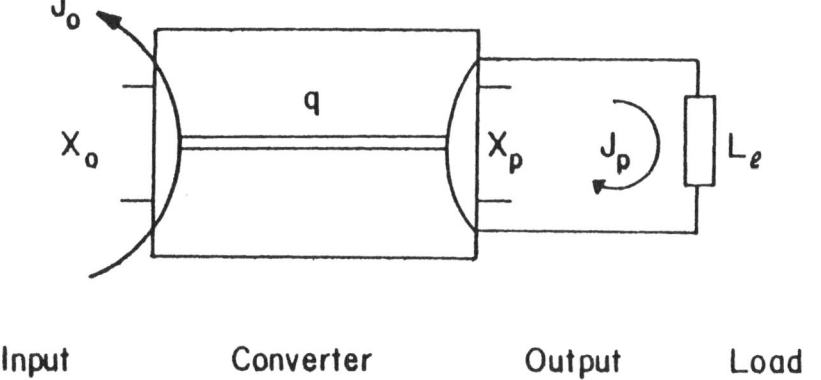

Input   Converter   Output   Load

Fig.1. Input-output scheme of oxidative phosphorylation.
Oxidative phosphorylation is depicted as a black box with an input
(oxidation) and an output (phosphorylation). The load conductance
L_ℓ summarizes the cellular ATP utilizing reactions. For the meaning
of the other symbols, see text.

with the input force X_o held constant (Stucki, 1980b). Note that
this system, in addition to linearity, also exhibits symmetry far
from equilibrium, i.e. at forces $X_i \gg RT$. The reason for the linearity of biological energy converters far from equilibrium
(Stucki et al, 1983) should not be confused with the conventional
Onsager type of linearity near equilibrium (Prigogine, 1967). The
former has to be understood as a result of an evolutionary design
to optimize efficiency whereas the latter is simply a result of
a small force approximation of the system near equilibrium (Stucki
et al, 1983).

Kedem and Caplan (1965) have defined a useful normalization of
the phenomenological coefficients L's : the degree of coupling q

(3)
$$q = L_{po}/\sqrt{L_p L_o}$$

and the phenomenological stoichiometry Z

(4)
$$Z = \sqrt{L_p/L_o}$$

Efficiency is defined as the ratio of output power and input power of the energy converter

(5) $$\eta = - J_p X_p / J_o X_o$$

Since oxidation is the spontaneous process and phosphorylation the driven process we have $X_o > 0$ and $X_p < 0$ with the flows being positive quantities. With these sign conventions η is confined to the interval $1 > \eta > 0$. This is a consequence of the second law of thermodynamics which requires that entropy production be positive definite:

(6) $$\dot{S}T = J_p X_p + J_o X_o > 0$$

The efficiency function has an interesting property when plotted versus the force ratio $x = ZX_p/X_o$ as shown in Figure 2. For any degree of coupling $q < 1$ efficiency passes through a maximum at a characteristic value of x. To be more specific we have that

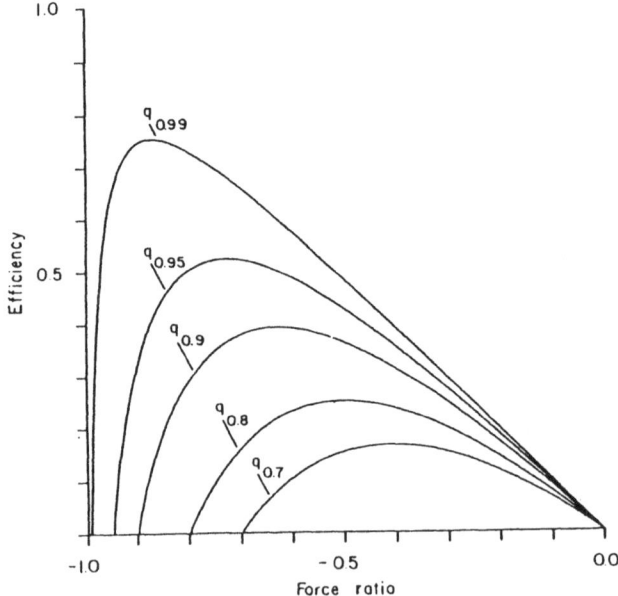

Fig.2. Efficiency of linear energy converters.
Plot of efficiency q versus the force ratio x according to equ.(5). Note that for each value of q this function displays an optimum η_{opt} at a unique value of the force ratio x_{opt} as defined in Equs. (7,8).

(7)
$$\eta_{opt} = \frac{q^2}{(1 + \sqrt{1 - q^2})^2}$$

and

(8)
$$x_{opt} = -\frac{q}{1 + \sqrt{1 - q^2}}$$

which are pure functions of q as shown by Kedem and Caplan (1965). It is reasonable to assume that in order to make the best use of the foodstuff, oxidative phosphorylation should always operate at η_{opt} for any given degree of coupling. The necessary and sufficient condition which allows the operation of the energy converter at optimal efficiency was found to be

(9)
$$L_\ell / L_p = \sqrt{1 - q^2}$$

whereby L_ℓ is the conductance of the load and L_p is the conductance of phosphorylation. This condition was therefore called conductance matching of oxidative phosphorylation (Stucki, 1980b).

We note in this relation q appears as a parameter. Hence the question arises as to what an appropriate value for q might be. Intuitively, one might assume that q = 1 would be the best choice because in a fully coupled energy converter there would be no loss of energy. A value of q = 1 implies, however, that L_ℓ = 0 in order to fulfill conductance matching. This means that the energy converter could operate at optimal efficiency only when there is no load attached to it ! Clearly this situation is of no biological interest. In fact, a fully coupled energy converter operating at optimal efficiency exactly corresponds to a Carnot machine operating with maximal efficiency at the state of thermodynamic equilibrium. For biological systems an equilibrium situation is identical to death. Therefore we must require that for biological energy converters q < 1. The exact value that q should assume depend entirely on the kind of output that is required from the energy converter. By making use of the transformation

(10)
$$\alpha = \arcsin (q)$$

one may define a parameter angle α for the degree of coupling. The different output functions can then be cast into the compact form of an output superfunction

(11) $\Omega = \tan^n(\alpha/2) \cos(\alpha)$

where the left hand side is normalized by the appropriate factors
(Stucki, 1982). Variation of the integer exponent n generates the
different output functions depicted in Figure 3. Thus for n = 1 we
obtain the maximal net flow of phosphorylation at optimal efficiency $(J_p)_{opt}$ at the value $q_f = 0.786$ (Stucki, 1980b). For n = 2 the
power $(J_p X_p)_{opt}$ is maximized at $q_p = 0.910$. The cases with n = 3
and n = 4 are the economic counterparts of the above outputs.
$(J_p\eta)_{opt}$ and $(J_p X_p\eta)_{opt}$ are maximized at $q_f^{ec} = 0.953$ and $q_p^{ec} = 0.972$
respectively. These latter two functions, in addition to maximize
flow or power simultaneously minimize the energy costs of energy
transformation (Stucki, 1980b; 1982). Note that all of these functions and associated maximizing values of the degree of coupling
have been found in the appropriate biological systems as is discussed
elsewhere (Stucki, 1982).

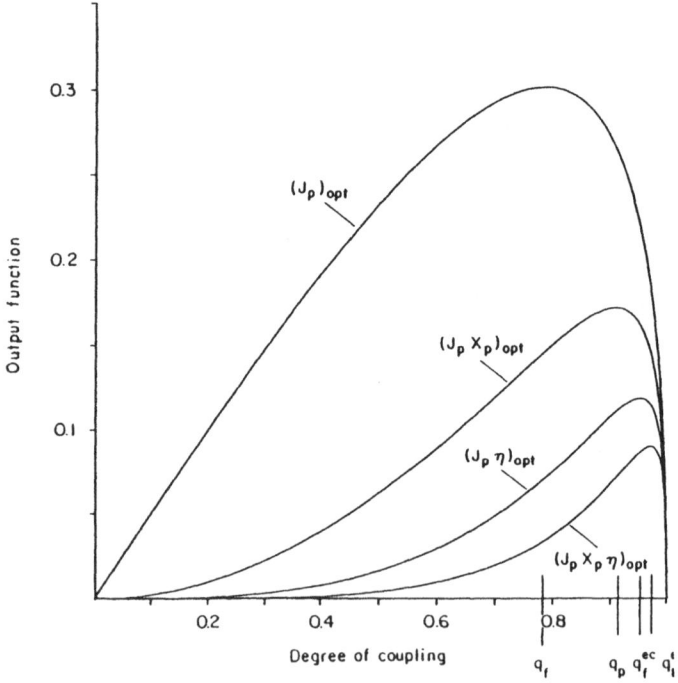

Fig. 3. Output functions of oxidative phosphorylation.
Plot of Ω versus q according to Equs. (10-11), for the values of
the exponent n = 1,2,3,4. For the physical interpretation and maximizing q's see text.

After having obtained an idea about possible values of q we are now faced with the next problem : conductance matching <u>in vivo</u>. Suppose that the system has selected an appropriate value of q and that this quantity is a constant for some time interval. Furthermore L_p can also be regarded as a constant since it is essentially composed of the rate constants of phosphorylation. Then in order to fulfill the condition of conductance matching also L_ϱ should be a constant. This is, however, almost certainly never the case. The load conductance in a living organism is typically a fluctuating quantity and at best the condition Equ. (9) can be expected to be fulfilled on a temporal average. Hence a fluctuating load conductance would constantly endanger conductance matching and the system could hardly ever operate at optimal efficiency.

Fortunately nature has found a way out of this dilemma : thermodynamic buffering. In oxidative phosphorylation this type of buffering is effected by the adenylate kinase reaction. Figure 4 shows how this reaction affects the adeninenucleotide concentrations. Whenever the load conductance is either higher or lower than the matched value, then the phosphate potential is driven away from its optimal value (Stucki, 1980a). By virtue of the reversible adenylate

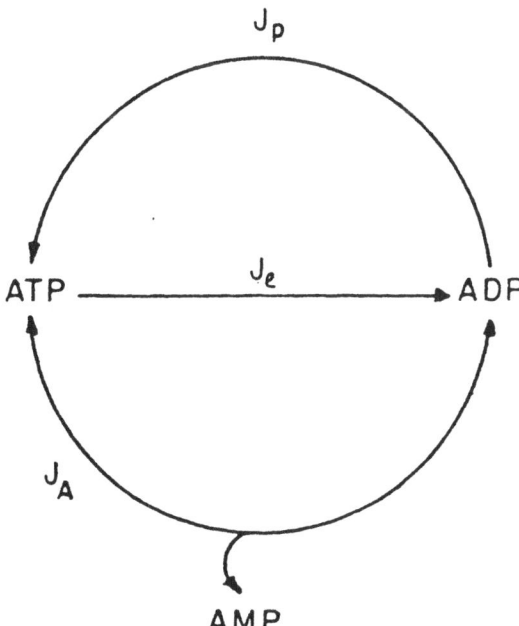

Fig.4. Effect of the flows on the adeninenucleotide concentrations. For the definitions of J's see text. This scheme defines the sign convention adopted for the differential equs. (15-17).

kinase reaction, however, this deviation can be minimized and de-
layed such that the phosphate potential stays close to the optimal
value even in the presence of a mismatched load. This feature is
illustrated in Figure 5 which shows the effects of a fluctuating
load on the force ratio and on efficiency. From this figure it
becomes quite clear that the adenylate kinase reaction can effectively damp the unwanted fluctuations of the force ratio and that
consequently efficiency remains always near its optimal value
(Stucki,1982; Stucki, 1980a).

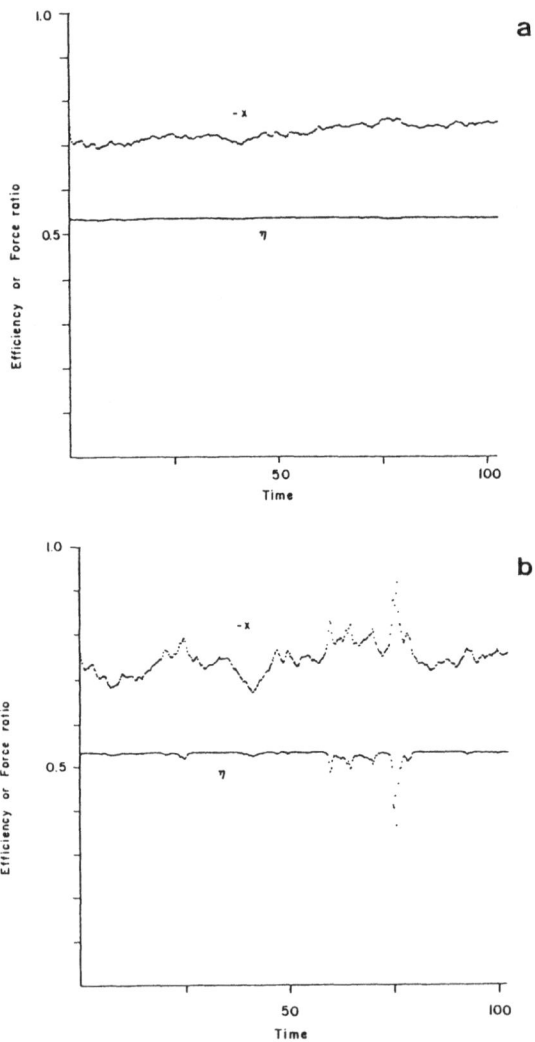

Fig.5. Fluctuating load and thermodynamic buffering.
Panel a : computer simulation of oxidative phosphorylation with a
fluctuating load conductance without adenylate kinase.
Panel b : same time history of the fluctuation as in panel a but
in the presence of adenylate kinase. For further details see
Stucki, 1980a.

It is now essential to realize that thermodynamic buffering is
a transient phenomenon. From Figure 4 it is evident that the adenylate kinase reaction can only manipulate X_p to remain close to an optimal value as long as there is either a net accumulation or a net utili-
zation of AMP, depending on whether the actual load conductance is
higher or lower than the matched value, respectively. Consequently
$dAMP/dt \neq 0$ during thermodynamic buffering. Since $\Sigma = ATP + ADP + AMP = const$
the same holds also for the time derivatives of the other adeninenucleotides. Therefore the system is not in a steady state as long as
thermodynamic buffering occurs. In other words, the system is in a
steady state if and only if the driving force of the adenylate kinase
reactions, the adenylate kinase potential

$$X_A = - \left[\Delta G_A^\circ + RT \ln (ATP \cdot AMP/ADP^2) \right] ,$$

vanishes. Therefore it was of interest to study the transient
kinetics of the thermodynamic buffering for the general cases
$X_A \neq 0$. Since the emphasis of my contribution is on energetics and
nonequilibrium thermodynamics I will now try to maintain a coherent description also for the treatment of transient phenomena.
Thus, in what follows, transient phenomena will be described in
terms of the nonequilibrium thermodynamic formalism rather than
switching now to the description of these processes in terms of
conventional kinetics.

TRANSIENT KINETICS OF THERMODYNAMIC BUFFERING

Before embarking into the detailed description of the transient kinetics of the adenylate kinase system we first chose a
convenient graphical representation of the transients. The space
of interest for our study is the positive orthant of the concentration space of the adeninenucleotides. The loci characterized by the
invariant $\Sigma = ATP + ADP + AMP$ consists in the intersection of this surface with R_+^3, namely in equilateral triangle. Since the surface Σ is
2-dimensional this triangle is a regular simplex. More precisely,
it is the reaction simplex of the adenylate kinase system (Stucki,
1984). Any steady state of the system can be represented as a
point in this simplex by using the following simple transformations

(12) $x = (ADP + (1/2) AMP)/\Sigma$

(13) $y = (\sqrt{3}/2) AMP / \Sigma$

with x and y being the image coordinates. Any path leading from one
steady state to another is a trajectory of the system. Note that
a trajectory is an integral curve and does not, therefore, contain
time information in an explicit manner. But the temporary evolution
of the system can be made visible by sampling and plotting successive points of the trajectory at constant time intervals. A long distance between adjacent points means a rapid evolution whereas

short distances correspond to a slow movement of the system from
one state to another.

There are a few characteristic lines in this simplex which are
of particular interest. Figure 6 depicts lines of constant adenylic
energy charge ε = (ATP +(1/2)ADP)/Σ, lines of constant phosphate
potential X_p = $- \left[\Delta G_p^o + RT \ln (ATP/ADP.P_i) \right]$ and lines of constant
adenylate kinase potential (see also Atkinson, 1977). According

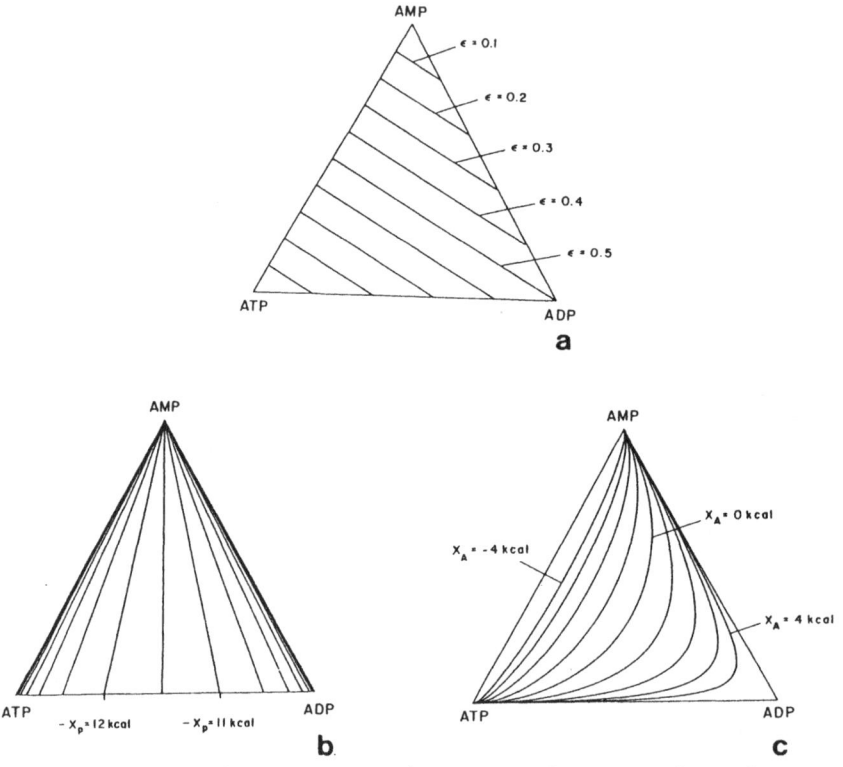

Fig.6. Different loci in the adeninenucleotide reaction simplex.
Panel a : lines of constant energy change ε; panel b : lines of
constant phosphate potentials X_p; panel c: lines of constant adenylate kinase potentials X_A. For definitions of ε, X_p and X_A see text.

to the remarks made at the end of the previous section it is clear
that any steady state of the system must lie on the equilibrium line
X_A = 0. This equilibrium line dissects the simplex in two regions.
In the region X_A > 0 we have thermodynamic buffering, i.e. a positive net flow of AMP due to

(14) $J_A = L_A X_A$

In contrast, within the region $X_A < 0$ the buffer is regenerated.

After this preamble we can now investigate the transients of oxidative phosphorylation plus the adenylate kinase reaction. For this end we first formulate the rate laws of the system. From Figure 4 we readily obtain the differential equations

$$(15) \qquad \dot{ATP} = J_p + J_\ell + J_A$$

$$(16) \qquad \dot{ADP} = -J_p - J_\ell - 2J_A$$

$$(17) \qquad \dot{AMP} = J_A$$

Note that one of these equations is redundant due to the invariant Σ, i.e. ATP + ADP + AMP = 0. By expressing now the flows J's as linear functions of the forces X's Equs. (15-17) can be written as

$$(18) \qquad \dot{ATP} = (L_p + L_\ell)X_p + L_{po}X_o + L_A X_A$$

$$(19) \qquad \dot{ADP} = -(L_p + L_\ell)X_p - L_{po}X_o - 2L_A X_A$$

$$(20) \qquad \dot{AMP} = L_A X_A$$

By introducing the definitions of X_p and X_A as functions of the adeninenucleotide concentrations, Equs. (18-20) constitute a set of autonomous nonlinear differential equations. We have solved these equations by numerical integration after introducing a convenient logarithmic transformation of the variables which is, however, not essential for our present purposes (Stucki, 1984).

Let us first study the evolution of the system from an arbitrary nonsteady initial state to the final steady state. The result of these simulations is depicted in Figure 7 for different activities of the adenylate kinase, i.e. for different values of L_A. Two main features of the system are apparent : first the kinetics shows a biphasic behaviour insofar as a rapid first phase is followed by a

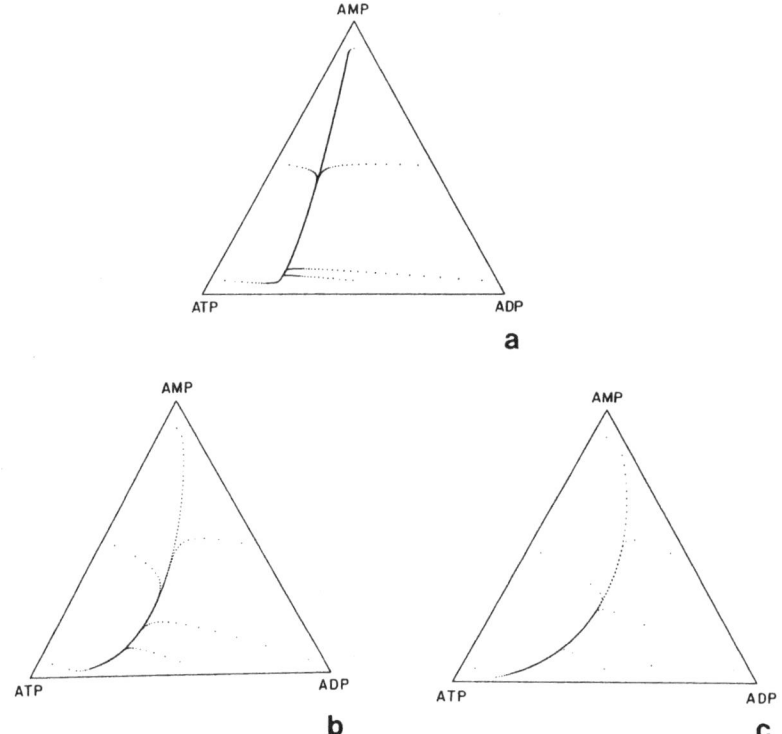

Fig.7.   Trajectories of oxidative phosphorylation plus adenylate
kinase.
The differential Equs. (18-20) were integrated numerically for dif-
ferent values of L_A and different initial adeninenucleotide concen-
trations. Panel a : L_A = 0.02; panel b : L_A = 0.2 , panel c: L_A = 1.0.
The degree of coupling was set to 0.953 and Z was assumed as 3.0.
For further details see Stucki (1984).
Sampling interval : 0.1 arbitrary time units of integration.

second slower phase of evolution of the trajectory to the steady
state. Second, all first phase parts of the trajectories lead to
a second phase part which is common to all trajectories. Later on,
I will demonstrate that this latter curve is a relaxation curve of
the system.

     An experimental verification of the trajectories is provided
by the results shown in Figure 8. In this experiment, isolated mitochondria from rat liver were incubated in the presence of different adeninenucleotide concentrations. Samples were then taken at
different time intervals and assayed for ATP, ADP and AMP. These
experimental data were plotted within the simplex by making use of
the transformations (12-13). With a nonlinear fit routine implemen-

ting the Marquardt algorithm (McIntosh and McIntosh, 1980) the
best fit of the theoretical curves based on Equs. (18-20) was
obtained by manipulating the parameters L_p, L_{po}, $L\ell$ and L_A. For
L_0 we assumed a constant value found in previous studies (Stucki,
1980). The comparison of the theoretical curves and the experimental data in Figure 8 demonstrates that the transient behaviour of
the system can be adequately described on the basis of the linear
relations Equs. (18-20). The reason for this linearity is discussed
elsewhere (Stucki, 1984).

Let us now try to understand these transients on a more general
basis. From Figure 7 it appears that at low activities of the adenylate kinase the first phase of the trajectories corresponds to an
evolution of the system at an almost vanishing AMP flow, i.e. along
a parallel to the baseline of the reaction simplex. The second phase
consists then in an evolution along a line of almost constant X_p
(see also Figure 6). On the other hand, at high activities of the
adenylate kinase, i.e. at large values of L_A, the situation looks
quite different. Here it appears that the first phase corresponds to
an evolution of the system along a line of constant adenylic energy
charge. The explanation for this observation is straightforward.
Whenever the conductance L_A is high then during the first phase the
flow J_A dominates all other flows. Hence, under these circumstances,
the system can be approximated by the simplified equations :

(21) $$\dot{ADP} = -2\,L_A X_A$$

(22) $$\dot{ATP} = +L_A X_A$$

Dividing the above equations yields the differential

(23) $$dADP/dATP = -2$$

which can be readily integrated to

(24) $$C = 2\,ATP + ADP$$

Defining the new constant

(25) $$\varepsilon = C\,/\,2\Sigma = (ATP + (1/2)ADP)/\Sigma$$

reveals that the adenylic energy charge is a constant of integration of the adenylate kinase reaction. We are now in the position
to state that the adenylic energy charge introduced by Atkinson on
intuitive grounds (Atkinson, 1977) has the physical interpretation
of a constant of motion of the adenylate kinase system. Therefore

it is now clear that the system must move along a line of constant ε
as long as all flows except J_A can essentially be neglected.

   The slow phase of the trajectories is somewhat more difficult
to explain. To fix ideas consider a typical valley such as the
Rhône valley in the Canton of Wallis in Switzerland. The profile of
this valley is such that water from the top of the mountains first
falls down a steep inclination to the bottom of the valley. This
water is then collected in the Rhône river which leads with a gentle
slope into the lake of Geneva. Intuitively the steep initial cre-
vasses correspond to the first phase of the trajectory, whereas the
gentle slide at the bottom of the valley corresponds to the second
part of the trajectories. How can we now translate this local geo-
graphic picture into thermodynamics ? The answer is , as mostly in
thermodynamics, entropy production and Lyapunov functions. At a

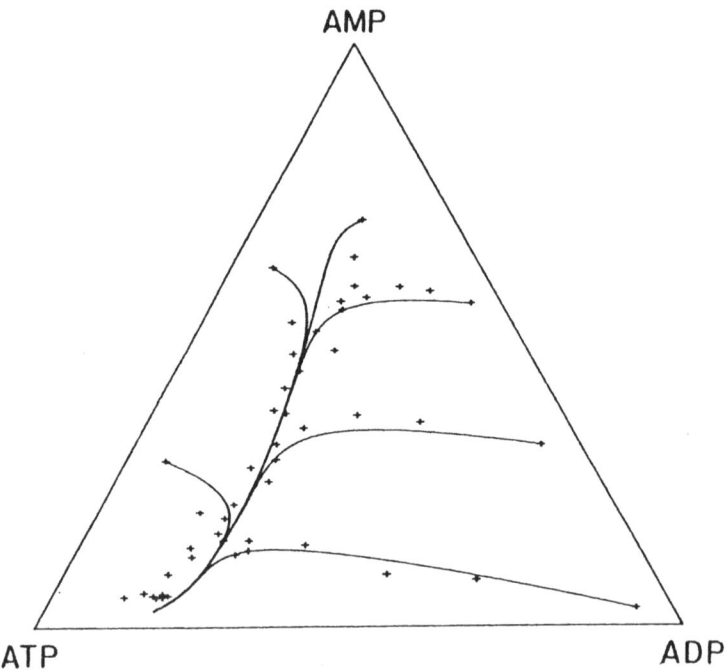

Fig.8. Experimental verification of trajectories.
The crosses depict measured adeninenucleotide concentrations determined in samples of incubated mitochondria from rat liver. Superimposed are the theoretical curves based on the differential equations
(15-17). The parameters L_p, L_{po}, L_ℓ and L_A were estimated with a
nonlinear fit routine using the Marquardt algorithm (McIntosh and
McIntosh, 1980). For further details see Stucki, 1984).

steady state the entropy production is minimal (Nicolis and Prigogine , 1977). Assume now that the system is just slightly displaced from the steady state. Again entropy production shows a local
minimum at that pseudo steady state. Hence the path of the river at
the bottom of the valley can be constructed by connecting the local
minima of a positive definite function. In other words, we have to
search for a scalar function which is positive definite and whose
time derivatives are negative definite everywhere in the reaction
simplex except at the steady state where it vanishes. Such a function is called a Lyapunov function (Stucki, 1978) and for linear
systems entropy production is always a Lyapunov function (Glansdorff
and Prigogine, 1971). The positive definitness of entropy production
for oxidative phosphorylation plus a load and the adenylate kinase
reaction is guaranteed by the second law of thermodynamics

$$(26) \qquad \dot{S}T = J_p X_p + J_\ell X_p + J_o X_o + J_A X_A \; > \; 0$$

The time derivative (Eulerian derivative) of this function assumes
the remarkably simple form

$$(27) \qquad \ddot{S}T = - \; 2RT \left[(\dot{ATP})^2/ATP + (\dot{ADP})^2/A\bar{DP} + (\dot{AMP})^2/AMP \right]$$

Hence entropy production is indeed a Lyapunov function. Note that
Equs. (26-27) constitute a sufficient proof for the global asymptotic stability of our system (Stucki, 1978). The temporal evolution
of entropy production can now be approximated by

$$(28) \qquad \dot{S}T = (t,\underset{\sim}{x}) = \; \dot{S}T \; (t0, \underset{\sim}{x0}) \; exp-(\lambda_{min}(t-t0))$$

where $\dot{S}$ depends now not only on the concentration vector of the
adeninenucleotides $\underset{\sim}{x}$ but also on time t. The exponent λ_{min} is
defined as (Stucki, 1984).

$$(29) \qquad\qquad\qquad \lambda_{min} = min \; (- \; \ddot{S}/\dot{S} \;)$$

Note that λ_{min} plays a similar role as do the eigenvalues for
linear systems of differential equations. Since the eigenvalues
are inversely related to the relaxation times of the system we can
infer that λ_{min} is related to the relaxation time governing the
long term behaviour of the system, i.e. the second phase of the
trajectories. Therefore by connecting the points in the simplex

corresponding to λ_{min} along the AMP coordinate we can construct
a path which approximates the second phase of the trajectories. For
the reasons stated above, I call such a path a relaxation curve of
the system.

Figure 9 depicts relaxation curves calculated for different
mismatched load conductances. A convenient parameter for a mismatched
load is defined by (Stucki, 1980)

(30)
$$L_\ell/L_p = \Theta\sqrt{1 - q^2}$$

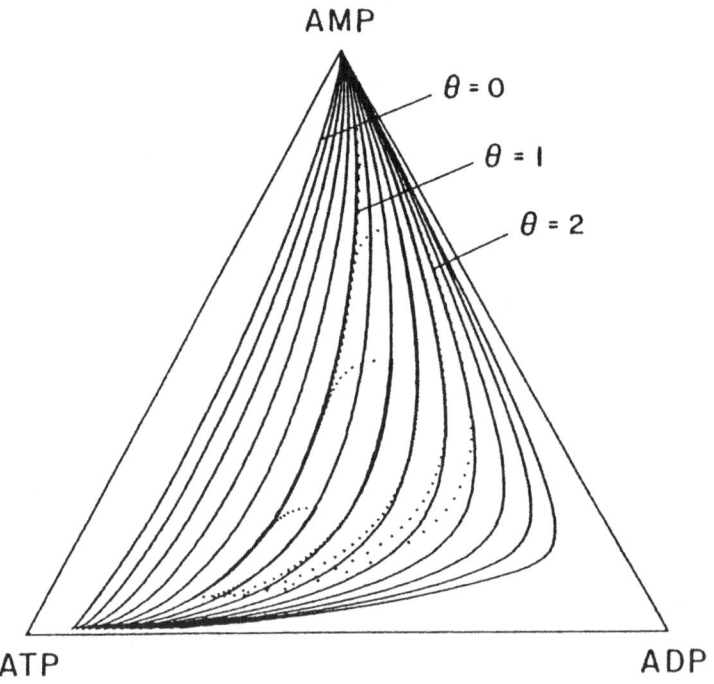

Fig.9. Relaxation curves.
Solid lines diplay the relaxation curves for various values of Θ
as defined in Equ. (30). Superimposed dotted curves are the solutions of the differential Equs.(15-17) during a work cycle as
described in the text. Sampling interval : 0.1 arbitrary time units
of integration. For further details see Stucki, 1984

If $\Theta = 1$ then we have conductance matching, else not. Θ is therefore a dimensionless measure for the deviation from conductance matching. Superimposed on these relaxation curves is the result of a computer simulation of Equs. (18-20). Initially in this simulation, Θ was set to a value $\Theta > 1$ as indicated in the Figure. The system was then allowed to evolve to the new steady state, i.e. until $X_A = 0$. After this Θ was switched back to 1 and the system evolved back to the original steady state. From the results depicted in Figure 9 it appears that the second slow phase of the trajectories can be very well approximated by these relaxation curves.

In summary we have now found an analytical function λ_{min} to approximate the second phase of the trajectories. This approximation thus obviates the necessity for the time consuming numerical integration of the differential equations (18-20). Most important, the λ_{min} are obtained from purely thermodynamic functions, namely entropy production and its time derivative. Hence it is indeed possible to approximate the long term kinetic behaviour of the system by thermodynamic functions. In addition, this formulation allows to maintain a coherent description of the system within the framework of nonequilibrium thermodynamics. I think that in the light of these results the choice of a formalism to deal with bioenergetic questions, kinetics or thermodynamics, is no longer merely a matter of "taste" as has been claimed (Reich and Sel'kov, 1981) but rather a matter of consistency of the basic definitions.

ACKNOWLEDGMENT

This work was supported by grants of Swiss National Science Foundation.

REFERENCES

Atkinson, D.E. (1977) Cellular Energy Metabolism and its Regulation, Academic Press , New-York.
Glansdorff, P. and Prigogine, I. (1971) Thermodynamic Theory of Structure, Stability and Fluctuations, Wiley, London.
Kedem, O. and Caplan, S.R. (1965)Trans. Farad. Soc. 21, 1897-1911.
McIntosh, J.E.A. and McIntosh, R.P. (1980) Mathematical Modelling and Computers in Endocrinology, Springer Verlag, Berlin.
Nicolis, G. and Prigogine, I. (1977) Self-Organization in Nonequilibrium Systems, Wiley, New-York.
Prigogine, I. (1967) Introduction to Thermodynamics of Irreversible Processes, John Wiley, New-York.
Reich, J.G. and Sel'kov, E.E. (1981) Energy Metabolism of the Cell, Academic Press, London.

Stucki, J.W. (1978) Prog. Biophys. Mol. Biol. 33, 99-187.
Stucki, J.W. (1980a) Eur. J. Biochem. 109, 257-267.
Stucki, J.W. (1980b) Eur. J. Biochem. 109, 269-283.
Stucki, J.W. (1982) in Metabolic Compartmentation (H. Sies, ed.)
Academic Press, London, pp. 39-69.
Stucki, J.W., Compiani, M. and Caplan, S.R. (1983) Biophys. Chem.
18, 101-109.
Stucki, J.W. (1984) Biophys. Chem. 19, 131-145.

SECTION VI - EVOLUTIONARY CONSIDERATIONS

EVOLUTION AND MUTATION OF THE AMINO ACID CODE

J. Tze-Fei Wong

Department of Biochemistry, University of Toronto

Toronto, Canada M5S 1A8

INTRODUCTION

The function of a protein is determined by the nature and sequence of its amino acid sidechains. Evolution of proteins must depend first of all upon development of a system of genetically encoded amino acids, and secondly upon optimization of their primary
sequences. Yet, in contrast to the extensive examination of evolution of protein sequences, the mechanisms by which the system of
proteinous amino acids had arrived at the present ensemble have
received only limited analysis. These mechanisms are the subject
of this study.

STARTING AMINO ACIDS

A serious impediment to the analysis of amino acid evolution
is posed by the uniformity of genetically encoded amino acids in
prokaryote, eukaryote, mitochondrial and chloroplast proteins.
Thus there is no reason at first glance to suppose that the ensemble had undergone any evolution at all. Broadly, there are five general theories regarding its evolution, which differ from each other
in terms of the number and nature of starting amino acids.

Frozen accident

According to Crick (1968), the choice of the 20 standard encoded amino acids might have stemmed from their chance admission
into the genetic code of an ancestral organism, and this membership of 20 has not been subject to evolutionary change ever since.

Doublet expansion

A triplet codon contains three letters. In the genetic code, a small amount of degeneracy is evident in the first letter, viz., UUR, CUR codons for Leu and CGR, AGR codons for Arg, but degeneracy involving the third letter is particularly widespread. Accordingly, the 64 codons may be regarded as 16 quartets. On this basis, Jukes (1966) proposed that the archetypal code functioned essentially as a doublet code, and coded for 15 amino acids plus the termination signal for translation. Subdivision of some of the quartets coupled to biosynthesis of novel amino acids later gave rise to the present 20.

Glycine metabolism

In the metabolic theory for expansion of the amino acid system proposed by Dillon (1973), the primordial organism began with only the single amino acid Gly. Evolving metabolism led to the generation of other amino acids from Gly, which spread through the genetic code. The postulated metabolic pathways for generating the other 19 amino acids, as in the less detailed formulation by Pelc (1965), are based on structural similarities between amino acids, and bear only incidental resemblance to the biosynthetic pathways of extant organisms.

Repeating triplets

The genetic coding sequences for proteins display a number of systematic characteristics, among which is an enrichment of RNY codons (R = purine, Y = pyrimidine, N = purine or pyrimidine). This has led Shepherd (1981) to postulate that the primordial mRNA is built up of repeating RNY triplets, coding for 8 amino acids. Subsequent elaboration of the code led to the present 20. On physical grounds, Eigen and Schuster (1979) also favoured repeating RNY triplets, and Crick et al. (1976) repeating RRY triplets.

Coevolutionary distribution

The coevolution theory for codon distribution began with the finding of overwhelming correlations between the biosynthetic relationships among amino acids and their codon allocations in the genetic code (Wong, 1975, 1976, 1980, 1981). To explain these correlations, it becomes necessary to postulate that most if not all biosynthetic product amino acids were absent initially, and received their codons from their precursor amino acids. This leaves 8 to 10 of the proteinous amino acids in the starting set.

Of these theories, the enrichment of RNY triplets in coding sequences may not be entirely historical in basis. The enrichment of R in the first letter position follows from the high frequency in proteins of amino acids with A or G in the first letter of their codons (Goel et ai., 1972). Besides, since the rate of synonymous substitution is as high as 5.1×10^{-9} per site per year (Jukes, 1980 ; Miyata et al., 1980), a primordial coding sequence consisting purely of repeating RNY units, unless maintained by selective pressure, would quickly lose its Y-enrichment at the third letter position. Accordingly the observed Y-enrichment may only be the result of functional selection, part of the specialized strategies adopted by different organisms with respect to codon utilisation (Grantham, 1980).

Frozen accident receives support from the lack of any observable variation among living organisms in the identities of the 20 encoded amino acids, which suggests that amino acid membership in the genetic code might be intrinsically immutable. Glycine metabolism dovetails with the Akabori mechanism for prebiotic polypeptide synthesis, which proceeds with formation of polyglycine followed by post-polymerization introduction of other amino acid sidechains (Akabori et al., 1956). However, both of these views are rigidly inconsistent with the results of prebiotic synthesis, which provide as possible starters much less than the standard 20 proteinous amino acids but much more than glycine.

Doublet expansion is supported by the "2 out of 3" reading mode of mitochondrial translation, which has been observed as well by Lagerkvist (1978) in ribosomal translation in vitro. However, it supplies no overall rationale to guide the subdivision of codon quartets between pairs of physically dissimilar amino acids, or to decide which member of a pair is ancestral. The prebiotic availability of both members of some of the pairs, and the sharing of the GAN quartet by Asp and Glu, both of which appear ancient, also pose difficulties.

Support for the coevolution theory is derived from four independent sources : correlation between codon allocation and biosynthetic precursor-product relationships, occurrence of pretranslational modifications, agreement between prebiotic unavailability and biosynthetic derivativeness, and mutability of the genetic code through competition.

COEVOLUTION THEORY

Many organisms, especially the prokaryotes, are proficient in the biosynthesis of amino acids through enzymic pathways that are constant for different organisms. The constancy suggests an ancient origin. In many of these biosynthetic pathways, one amino acid serves as precursor to another amino acid. When the codons for the product amino acids are compared to the codons for their precursors, the former are found to display a striking contiguity (viz., separation by only a single base change), with the latter.

The aggregate probability of finding so many contiguities between the codons of precursors and of products is less than 2×10^{-4}, counting only the Ser-Trp, Ser-Cys, Val-Leu, Thr-Ile, Gln-His, Phe-Tyr, Glu-Gln and Asp-Asn precursor-product pairs in Table 1. In addition, the correlation between codon allocation and biosynthesis is strengthened by a nonrandom concentration of the biosynthetically related pairs Ile-Met, His-Gln, Asn-Lys and Cys-Trp among those sharing codon quartets (Wong, 1975).

Since amino acid biosynthesis is not functionally linked to the genetic code, these observed correlations are in all likelihood historical in origin. They suggest that the code coevolved with the biosynthetic pathways, and the product amino acids received codons that had belonged to their respective precursors. The transfer of codons from precursor to product could be achieved by one of two mechanisms. If the product resembled the precursor physically, it could compete for attachment to the tRNA adaptors utilised by the precursor. If there was little resemblance, as with the Ser-Trp pair, a pretranslational modification of the precursor while attached to its tRNA adaptor could bring about the transfer. Transfer by either mechanism would enable the product to receive part or all of the codons of the precursor, assuring contiguities between precursor and product codon domains.

PRETRANSLATIONAL MODIFICATION

Two instances of pretranslational modifications are known among living organisms :

$$Met-tRNA^{Met} \longrightarrow Formyl-Met-tRNA^{Met}$$
$$Glu-tRNA^{Gln} \longrightarrow Gln-tRNA^{Gln}$$

By the pretranslational formylation of Met-tRNA^{Met}, fMet is incorporated into the N-terminal position of prokaryotic proteins. Similarly, in Gram-positive bacteria, Glu is first attached to both tRNA^{Glu} and tRNA^{Gln}. The Glu-tRNA^{Glu} does not undergo further transformation, but the Glu-tRNA^{Gln} is amidated by free Gln or Asn to form Gln-tRNA^{Gln} before translation at the ribosome to incorporate Gln into nascent proteins (Wilcox and Ninrenberg, 1968). Lapointe and coworkers found no glutaminyl-tRNA synthetase in B. subtilis to attach Gln directly to tRNA^{Gln}, and a single glutamyl-tRNA synthetase to attach Glu to both tRNA^{Glu} and tRNA^{Gln} (Kern et al., 1979 ; Lapointe, 1982 ; Proulx et al., 1983).

This pathway of Gln incorporation by way of pretranslational modification takes place in B. subtilis, even though free Gln is readily available ; these cells contain glutamine synthetase evidently in excess of 1 % of total cell protein (Deuel et al., 1970). The one plausible explanation is that, in keeping with the coevolu-

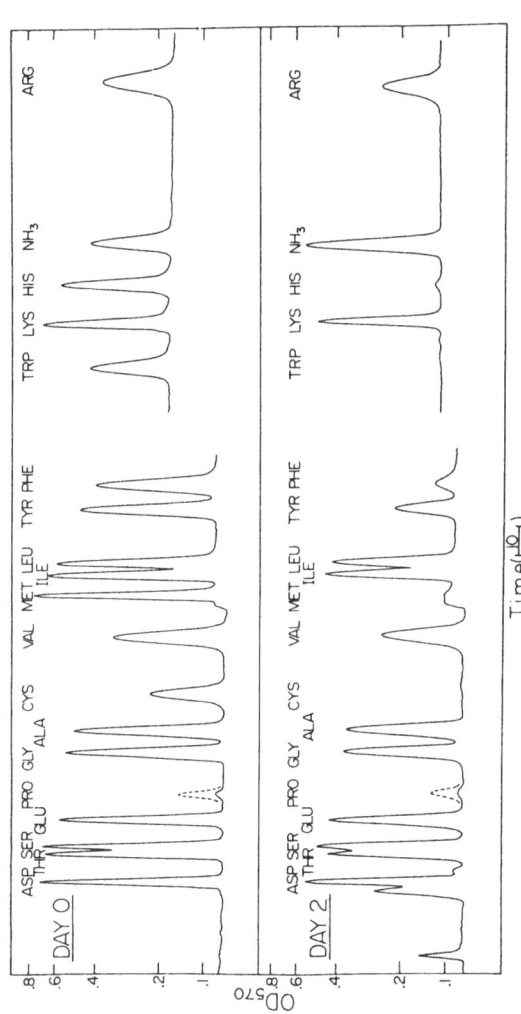

Figure 1. Ultraviolet degradation of amino acids. Protein hydrolysate standard containing 17 standard amino acids plus cystine and ammonium sulfate was supplemented with Trp and exposed to General Electric G1518 15W germicidal lamp in 0.1 M sodium phosphate buffer, pH 7 at a distance of 5.5 cm for 0 and 48 hours. Energy flux was estimated with a Blak-Ray J225 UV Meter (Ultraviolet Products, San Gabriel, California) to be 1.8 mW/cm² at that distance. Analysis of amino acids was performed on a Beckman 121 Analyzer. DC1A resin (0.9 x 56 cm) was employed for separation of acidic and neutral amino acids, with 0.2 N sodium citrate at pH 3.25 (120 minutes) and pH 4.25 (100 minutes). PA35 resin (0.9 x 18 cm) was employed for basic amino acids, with 0.35 N sodium citrate, pH 5.25. Dotted line shows absorbance for Pro at 440 nm.

tion theory, originally both tRNA^{Glu} responding to the GAR codons, and tRNA^{Gln} responding to the CAR codons, belonged to Glu. When primitive catalysis amidated Glu-tRNA^{Gln} to form Gln-tRNA^{Gln}, Gln would take over the CAR codons from Glu. This pathway for Gln incorporation no longer operates in E. Coli or in eukaryotes.

PREBIOTIC SYNTHESIS

        Beginning with Miller, it is known that amino acids are readily synthesized under simulated prebiotic conditions (Miller, 1953 ; Miller et al., 1976). A wide range of energy inputs may be used, including electric discharge, shock wave, heat and high-energy radiation. A survey of the actual amino acids obtainable from such prebiotic synthesis has revealed two important kinds of discrepancies between the prebiotic amino acids and the proteinous amino acids encoded by the genetic code : (a) Inadequate synthesis – Many of the proteinous amino acids are not produced in any significant quantity prebiotically, and (b) Incomplete encoding – many amino acids produced prebiotically are not encoded by the present-day genetic code.

        A possible explanation for these discrepancies is that the experimental conditions employed for the prebiotic synthesis represent incorrect models for conditions on primitive earth in terms of either starting materials or the form of energy input. If this were the case, improvements in experimental design might eventually lead to the synthesis of all of, and only the encoded amino acids. However, such an explanation is rendered unlikely by the close resemblance between the profile of amino acids obtainable from electric discharge, and the profile found in carbonaceous meteorites which suggests that electric discharge provides a useful model at least for formation of the amino acids on meteorites, if not on primordial earth as well.

        The more complex proteinous amino acids, besides being difficult to synthesize prebiotically, are also unstable to thermal and ultraviolet degradation. Gln and Asn are thermally too unstable to accumulate on primitive earth (Wong and Bronskill, 1979). Lys, Arg, Cys, Trp, His and Met also exhibit lower activation energies for thermal decomposition than other amino acids (Olafsson, 1971). In addition, the absence of a prominent ozone screen from stratosphere of primitive earth would expose the oceans to intense ultraviolet radiation from the sun (Sagan, 1965). As the observations recorded in Fig. 1 indicate, Cys, Met, Trp, His, Tyr and Phe would be subject to substantial photodegradation, in accord with the known photolabilities of these complex amino acids (McLaren and Shugar, 1964).

        Thus the amino acids Asn, Gln, Trp, Arg, His and to a lesser

extent Phe, and Tyr, suffer the twin disadvantages under prebiotic
conditions of difficult synthesis and easy degradation. Therefore
their availability at the start of biological evolution is entirely
questionable. Although Met and Cys have been obtained under specialized synthetic conditions (Steinman et al., 1968 ; Sagan and Khare,
1971), their limited stabilities and the single codon for Met point
to late arrival as well. Most of these unlikely starters share a
common attribute : nowadays they are biosynthetically derived from
other amino acids (Table 1). This correlation between deficient
prebiotic synthesis and biosynthetic derivativeness suggests that
many of the amino acid biosynthetic pathways in fact closely resemble the pathways by which prebiotically unavailable amino acids
first gained entry into the evolving biosphere. It confirms the
coevolution theory that only about half of the 20 encoded amino
acids of today were among the set of starting amino acids that
made up primordial proteins. The set at one time or another likely
included such amino acids as norvaline, isovaline and α -aminobu-
tyric acid, which had been available prebiotically, but long since
rejected from the evolving code.

MUTATION OF GENETIC CODE

The entry of one group of amino acids such as Asn, Gln and
Trp into the genetic code, and exit of another group such as norvaline and isovaline, added up to an actively evolving code. Since
the protein sequences of an organism are optimised, or close to
being optimised, with respect to its encoded amino acids, complete
or partial replacement of any incumbent encoded amino acid by a
newcomer must act against the preexistent optimisation. This initial disadvantage constitutes a selective barrier that must be overcome even if the introduction of the novel amino acid offers real
advantages in the long term. The strength of this barrier is attes-
ted to by the apparent immutability of the amino acid code throughout the biospere.

Because of the apparent immutability, it becomes necessary
to consider what kind of mutational pathways could possibly bring
about a membership change back in the coevolutionary age. Fig. 2
illustrates plausible pathways capable of achieving the postulated
displacement of homoserine, or Hse, by its biosynthetic product
Thr. Such displacement requires removal of old sequence optimisation favouring Hse over Thr, and establishment of new sequence optimisation favouring Thr over Hse.

In pathways I, AA_g represents a protein residue that could be
occupied by any amino acid. A Hse-residue functions well only when
occupied by Hse, but not when occupied by Thr. Likewise, a Thr-
residue functions well only with Thr but not with Hse, whereas a
Hse/Thr-residue functions well with either Hse or Thr. Mutation of

Table 1. Prebiotic and biological synthesis of amino acids.

Amino Acid	Membership in genetic code	Presence on Murchison meteorite	Synthesis by electric discharge	Biosynthesis	
				Amino Acid precursor	Enzymic pathway
Gly	+	1.0	1.0	–	
Ala	+	0.54	1.8	–	
Ser	+	–	0.01	–	
Asp	+	0.09	0.08	–	
Glu	+	0.49	0.02	–	
Val	+	0.23	0.04	–	
Pro	+	0.24	0.003	Glu	pyrroline carboxylate reductase
Leu	+	–	0.03	Val	α-isopropylmalate synthetase
Thr	+	–	0.002	Asp	Hse dehydrogenase
Ile	+	–	0.01	Thr	Thr deaminase
Phe	+	–	–	–	
Tyr	+	–	–	Phe	Phe-4-monooxygenase
Cys	+	–	–	Ser	O-acetyl-Ser thiol-lyase
Lys	+	–	–	Asp	2,6-diaminopimelate pathway
Arg	+	–	–	Glu	urea cycle
His	+	–	–	Gln	amidotransferase
Met	+	–	–	Thr/Hse	O-succinyl-Hse thiol-lyase
Trp	+	–	–	Ser	Trp synthase
Asn	+	–	–	Asp	Asn synthetase
Gln	+	–	–	Glu	Gln synthetase
β-Alanine	–	0.08	0.04		
Sarcosine	–	0.07	0.13		
N-Ethylglycine	–	0.07	0.07		
N-Methylalanine	–	0.02	0.03		
Norvaline	–	0.08	0.14		
Isovaline	–	0.17	0.01		
α-Amino-n-butyric acid	–	0.26	0.61		
α-Amino-isobutyric acid	–	0.56	0.07		
β-Amino-n-butyric acid	–	0.11	0.001		
β-Amino-isobutyric acid	–	0.06	0.001		
γ-Amino-butyric acid	–	0.04	0.005		
Pipecolic acid	–	0.14	0.0001		

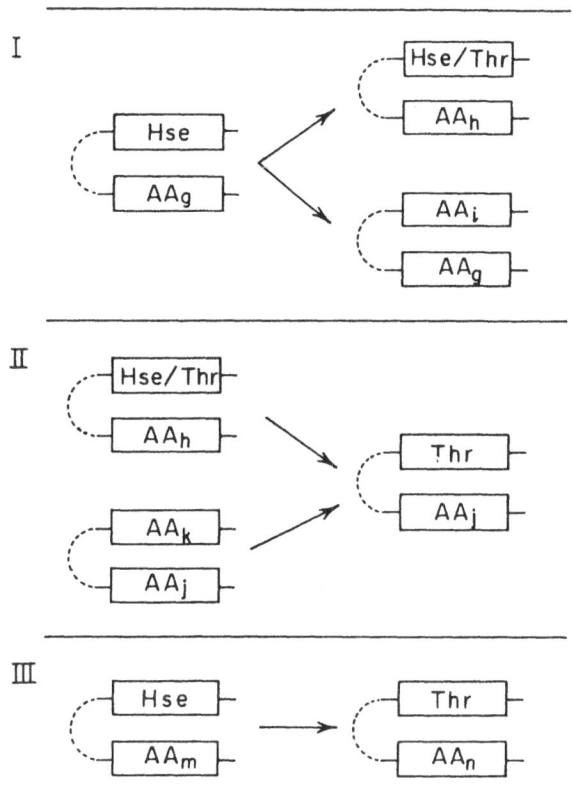

Figure 2. Mutational pathways for displacement of Hse by Trp. Each pair of amino acids represent residues, on the same or neighbouring polypeptide chains, that must interact cooperatively for proper protein function.

AA$_g$ to AA$_h$ alters the environment of the Hse-residue nearby, and converting it into a Hse/Thr-residue ; AA$_g$ but not AA$_h$ specifically demands Hse in its vicinity for proper function. A specific requirement for Hse as opposed to Thr also may be eliminated when the Hse-residue mutates to some other kind of amino acid AA$_i$. In pathways II, mutation of AA$_h$ to AA$_j$ further modulates the environment of the Hse/Thr residue, converting it into a Thr-residue. A Thr-residue also may be created when AA$_k$ mutates to homoserine. AA$_j$ cooperates well with AA$_k$ or Thr, but not with Hse, in its immediate vicinity.

In pathways I and II, the removal of sequence bias favouring Hse and the establishment of sequence bias favouring Thr occur in two separate steps. In pathway III, these steps are combined, and

the single mutation of AA_m to AA_n converts a Hse-residue directly to
a Thr-residue.

The recognition of plausible pathways for mutating the membership of the genetic code implies that the code, appearance of perma-
nence and frozen accident to the contrary, is potentially mutable.
It remains essential, however, to seek an experimental verification
of this mutability. Accordingly, we have sought to replace the incumbent Trp in the genetic code of B. subtilis by its fluorinated

Table 2. Serial mutation in successive stages of Bacillus subilis
strain QB928 from Trp preference to 4-FTrp preference for growth
(Wong, 1983)

Stage	Strain	Growth in Trp / Growth in 4-FTrp
0	QB928	730
1	LC8	360
2	LC33	1.7
3	HR7	0.66
4	HR15	0.036

analogue 4-fluorotryptophan, or 4-FTrp. As Table 2 shows, the B.
subtilis Trp-auxotrophic chain QB928 began with a large Trp/4-FTrp
growth ratio, which reflects a sequence optimisation decisively in
favour of Trp over 4-FTrp. Upon serial mutations of QB928 to successively LC8, LC33, HR7 and finally HR15, the Trp/4-FTrp growth
ratio decreases to far below unity, which points to a sequence optimisation decisively in favour of 4-FTrp over Trp. Due to this rever-
sal in the relative fitness of Trp and 4-FTrp, 4-FTrp has replaced
Trp as a competent member of the genetic code for the support of
cell growth. Although the sites of the serial mutations from
QB928 to HR15 have yet to be elucidated, these mutations establish
that the genetic code is by no means immutable.

Thus, starting from prebiotically available amino acids,
co-evolution with amino acid biosynthesis led to the emergence of
the 20-member amino acid code. Given its new-found mutability,
this code, despite over two billion years of invariance, has yet
to become a closed book.

ACKNOWLEDGEMENTS

       This study has been supported by the Medical Research Council
of Canada.

REFERENCES

Akabori, S., Okawa, K. and Sato, M. (1956) Bull. Chem. Soc. (Japan)
29, 608.
Crick, F.H.C. (1968) J. Mol. Biol. 38, 367-379.
Crick, F.H.C., Brenner, S., Klug, A. and Pieczenik, G. (1976) Ori-
gins of Life 7, 389-397.
Deuel, T.F., Ginsburg, A., Yeh, J., Shelton, E. and Stadtman, E.R.
(1970) J. Biol. Chem. 245, 5195-5205.
Dillon, L.S. (1973) Bot. Rev. 39, 301-345.
Eigen, M. and Schuster, P. (1979) "The Hypercycle", Springer-Verlag,
Berlin, Heidelberg, New-York, p. 64.
Goel, N., Rao, G.S. and Ycas, M. (1972) J. Theor. Biol. 35, 399-
457.
Grantham, R. (1980) Trends Biochem. Sci. 5, 327-331.
Jukes, T.H. (1966) "Molecules and Evolution", Columbia University
Press, New-York and London, pp. 65-70.
Jukes, T.H. (1980) Naturwissenschaften 67, 534-539.
Kern, D., Potier, S., Boulanger, Y. and Lapointe, J. (1979) J. Biol.
Chem. 254, 518-524.
Lagerkvist, U. (1978) Proc. Natl. Acad. Sci. USA 75, 1759-1762.
Lapointe, J. (1982) Can. J. Biochem. 60, 471-474.
McLaren, A.D. and Shugar, D. (1964) "Photochemistry of Proteins and
Nucleic Acids" McMillan Co., New-York, p. 97.
Miller, S.L. (1953) Science 117, 528-529.
Miller, S.L., Urey, H.C. and Oro, J. (1976) J. Mol. Evol. 9, 59-72.
Miyata, T., Yasunaga, T. and Nishida, T. (1980) Proc. Natl. Acad.
Sci. USA 77, 7328-7332.
Olafsson, P.G. and Bryan, A.M. (1971) Polymer Letters 9, 521.
Pelc, S.R. (1965) Nature 207, 597-599.
Proulx, M., Duplain, L., Lacoste, L., Yagushi, M. and Lapointe, J.
(1983) J. Biol. Chem. 258, 753-759.
Ring, D., Wolman, Y., Friedmann, N. and Miller, S.L. (1972) Proc.
Natl. Acad. Sci. USA 69, 765-768.
Sagan, C. (1965) in "The Origins of Prebiological System", S.W. Fox,
ed.,Academic Press, New-York, London, p. 238-239.
Sagan, C. and Khare, B.N. (1971) Science 173, 417-420.
Shepherd, J.C.W. (1981) J. Mol. Evol. 17, 94-102.
Steinman, G., Smith, A.E. and Silver, J.J. (1968) Science 159,
1108-1109.
Wilcox, M. and Nirenberg, M. (1968) Proc. Natl. Acad. Sci. USA 61,
229-236.
Wolman, Y., Haverland, W.J. and Miller, S.L. (1972) Proc. Natl.
Acad. Sci. USA 69, 809-811.

Wong, J.T. (1975) Proc. Natl. Acad. Sci. USA 72, 1909–1912.
Wong, J.T. (1976) Proc. Natl. Acad. Sci. USA 73, 2336–2340.
Wong, J.T. (1980) Proc. Natl. Acad. Sci. USA 77, 1083–1086.
Wong, J.T. (1981) Trends Biochem. Sci. 6, 33–36.
Wong, J.T. (1983) Proc. Natl. Acad. Sci. USA 80, 6303–6306.
Wong, J.T. and Bronskill, P.M. (1979) J. Mol. Evol. 13, 115–125.

FROM BINDING TO CATALYSIS - INVESTIGATIONS USING SYNTHETIC PEPTIDES

Bernd Gutte, Rudolf Moser, Stephan Klauser
and Martin Weilenmann

Biochemisches Institut der Universität Zürich
CH-8057 Zürich, Switzerland

INTRODUCTION

Receptors, repressors, protein inhibitors, and antibodies
are proteins that bind a ligand but do not alter it chemically.
Receptors act by forming membrane channels (acetylcholine receptor),
by activating membrane-bound enzymes (adrenaline receptor), or by
regulating gene expression (nuclear estrogen receptor). Repressors
can prevent transcription of bacterial operons by binding very
tightly to the operator region of the operon. Repressor binding is
controlled allosterically by metabolites. Protein inhibitors almost
irreversibly inactivate enzymes by forming extremely stable enzyme-
inhibitor complexes ; dissociation constants of the order of $10^{-13}M$
have been found. Antibodies bind and neutralize bacteria, viruses,
toxins etc., and thus protect the organism against disease.

Enzymes are proteins that bind and chemically convert a ligand. For the purpose of this discussion they are divided into two
groups. Group I enzymes (e.g. ribonuclease A, lysozyme, chymotryp-
sin) solely require "correctly" folded polypeptide chains to be
biologically active ; group II enzymes (e.g. carboxypeptidase A,
glyceraldehyde-3-phosphate dehydrogenase, pyruvate dehydrogenase)
in addition need metal ions and prosthetic groups for activity. Enzymes exert their catalytic power by lowering the activation energy
of a reaction. The activated complexes that are formed between enzymes and substrates can be covalent or non-covalent. In covalent
complexes the substrate is bound chemically either to the apoenzy-
me as in glyceraldehyde-3-phosphate dehydrogenase or to prosthetic
groups as in pyruvate dehydrogenase.

The distinction between ligand-binding proteins and enzymes raises the question of their interconvertibility. There are many examples of enzymes that have become inactive through mutation or chemical modification of essential residues but have retained the ability to bind the substrate (Fig. 1, step 4). Extensive studies on this subject have been performed with the ribonuclease S-pepti-de-S-protein system (Richards and Wyckoff, 1971). The reverse step (Fig. 1, step 3) must have occurred in nature many times during the billions of years of evolution but so far has not been achieved in the laboratory apart from the reversible modification of catalytically essential residues (Means and Feeney, 1971).

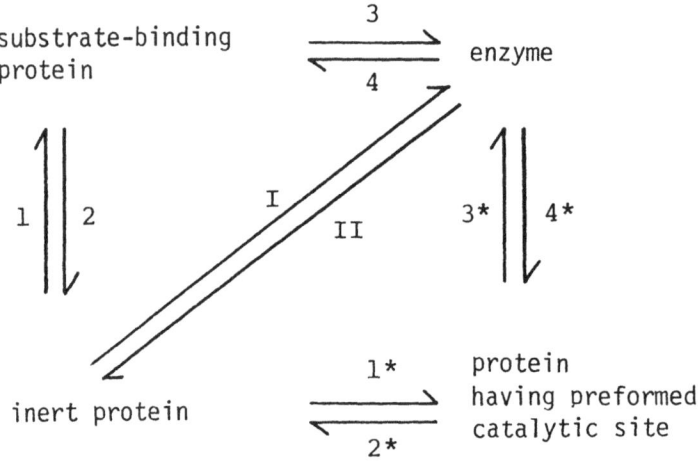

Figure 1. Possible interconversions between inert protein and enzyme. Steps 1* and 3, acquirement of catalytic site ; steps 1 and 3*, acquirement of substrate binding site.

The conversion of trypsinogen to trypsin can be considered a model reaction for step 3* of Fig. 1. The geometry of the charge relay system (i.e. the active site) of zymogen and enzyme is nearly identical ; the major conformational change occurring during trypsin activation is the formation of the substrate specificity pocket (Huber and Bode, 1978). It seems that this step can also be performed in the test tube. An artificial 34-residue polypeptide had weak ribonuclease activity with high preference for cleavage at the 3'-end of C when the alternating positions 1, 3 and 5 were occupied by the aromatic residues Phe, Phe and Tyr, respectively

(Gutte et al., 1979). Replacement of these residues by Val, Val and Ile, respectively, abolished nucleotide binding almost completely whereas the conformation of the residues involved in catalysis was most likely not affected (Gutte, unpublished results).

The direct transition from an inert protein to an enzyme or vice versa without the formation of an intermediate (Fig. 1, steps I and II) cannot be excluded.

SYNTHESIS OF MODEL PEPTIDES POSSESSING BINDING SPECIFICITY

Based on the known geometry or topology of a binding site, a number of model peptides possessing specific binding properties of various natural proteins has been prepared.

"Surface Simulation Synthesis" (Lee and Atassi, 1976) of an Antigenic Determinant of Lysozyme

Most antigenic determinants of proteins are formed by amino acid residues that are adjacent in the tertiary structure but distant in the sequence. Linking these residues either directly or via spacer amino acids, Lee and Atassi (1976) synthesized a linear decapeptide indistinguishable in its immunological properties from the corresponding antigenic determinant of native lysozyme.

A Helical 22-Residue Model of Plasma Apolipoprotein A-I

Fukushima et al. (1979) designed and synthesized an amphiphilic helical 22-residue polypeptide which has the surface properties of plasma apolipoprotein A-I. The hydrophilic side of the helical cylinder contained mainly glutamic acid and lysine residues and interacted with water, the hydrophobic side was formed by leucine residues and bound to phospholipids. The homology of this peptide with residues 198-219 of human apolipoprotein A-I representing one of the six highly homologous 22-residue segments with strong helix-forming potential was 41 %.

A Cyclic Octapeptide Model of the Zn(II)-Binding Site of Carboxypeptidase A

Iyer et al. (1981) designed and synthesized a cyclic octapeptide (Fig. 2a) mimicking the Zn(II)-binding site of carboxypeptidase A. In carboxypeptidase A, the essential Zn(II) is coordinated to one glutamic acid and two histidine residues, the fourth ligand in the absence of substrate being water (Blow and Steitz, 1970). The synthetic model contained also one glutamic acid and two histidine residues separated by glycine spacers. The sequence of these residues was deduced from model building. ^{1}H- and ^{13}C-NMR studies showed that Zn(II) was ligated to the side chains of the two histidine residues and the glutamic acid residue of the designed cyclic

a.   cyclo-(Gly-Glu-Gly-Gly-His-Gly-His-Gly)

b.   cyclo-(Pro-Phe7-D-Trp8-Lys9-Thr10-Phe11)

```
                  1          5              10                15
c.   native sequence:  Lys-Glu-Thr-Ala-Ala-Ala-Lys-Phe-Glu-Arg-Gln-His-Met-Asp-Ser


                  1          5              10                15
     model sequence:  Ala-Glu-Ala-Ala-Ala-Ala-Lys-Phe-Ala-Arg-Ala-His-Met-Ala-Ala
```

d.

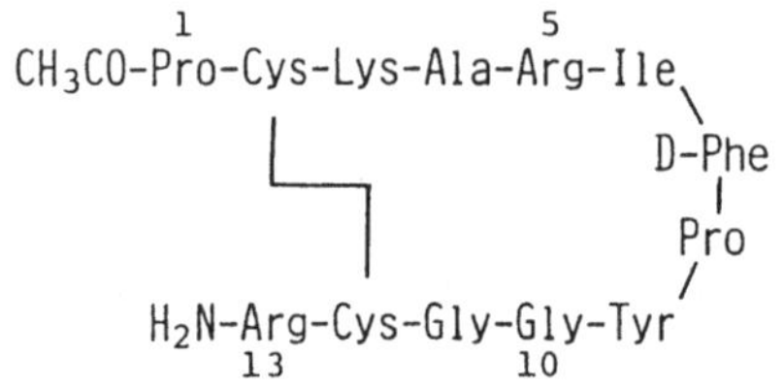

Figure 2.
a. Cyclic octapeptide model of the Zn(II)-binding site of carboxypeptidase A.
b. Potent cyclic hexapeptide analogue of somatostatin. Numbers indicate the positions of the corresponding residues in the natural hormone. In somatostatin, residue 8 is L-tryptophan.
c. NH₂-terminal 15-residue fragment of ribonuclease A (RNase A
1-15, native sequence) and synthetic analogue of the fragment
(model sequence).
d. A 13-residue trypsin inhibitor resembling the binding site of
BPTI. The segments Pro 1 to Ile 6 and Tyr 9 to Arg 13 of the synthetic model are identical with segments Pro 13 to Ile 18 and
Tyr 35 to Arg 39 of the natural inhibitor, respectively.


```
           1               5                    10
     Glu-Phe-Ala-Ala-Glu-Glu-Ala-Ala-Ser-Phe
```

Figure 3. Sequence of a designed decapeptide with glycosidase
activity.

octapeptide. In linear form, the octapeptide did not have significant Zn-binding specificity.

A Potent Cyclic Hexapeptide Analogue of Somatostatin

Conformational analysis and the use of a computer modelling
and graphics system led to the design and synthesis of a highly
active cyclic hexapeptide analogue of the 14-residue peptide hormone somatostatin (Fig. 2b) (Veber et al., 1981). In this analogue,
nine of the 14 residues of somatostatin were replaced by a single
proline. The resulting conformational constraint was most likely
responsible for the high activity of the analogue. It also reduced
the susceptibility to digestion by proteolytic enzymes and thus
allowed prolonged duration of action. These properties render this
kind of analogue highly interesting in tests of the potential application of somatostatin-like compounds in diabetes therapy.

A Simplified Functional Analogue of Ribonuclease S-Peptide

Komoriya and Chaiken (1982) synthesized a 15-residue analogue
of ribonuclease S-peptide (Richards and Wyckoff, 1971) predicted
to have the essential sequence information needed to produce a stable and enzymatically active noncovalent complex with ribonuclease
S-protein. In the synthetic analogue, six residues of the natural
sequence were replaced by alanine residues (Fig. 2c). It was found
that the model peptide indeed formed a stable complex with ribonuclease S-protein which could be crystallized. The crystals were
isomorphous with those of the complex consisting of natural S-peptide and natural S-protein.

A 13-Residue Peptide Resembling the Primary Binding Site of the Basic Pancreatic Trypsin Inhibitor (BPTI)

The X-ray structure of the trypsin-inhibitor complex (Huber
et al., 1974) shows that those residues of BPTI that are in van der
Waals contact with trypsin residues are arranged sequentially on
two antiparallel β-strands (legend of Fig. 2d). Kitchell and Dyckes
(1982) synthesized a 13-residue model inhibitor (Fig. 2d) which comprised the trypsin-binding region of the natural inhibitor and in
which residues 19 to 34 of BPTI were replaced by -D-Phe-Pro-. This
dipeptide unit has the potential to form a tight β-turn that would
allow antiparallel alignment of residues 1-6 with residues 9-13.
The resulting structure of the model inhibitor would resemble closely the structure of the trypsin binding site of BPTI.

The cyclic 13-residue cystine peptide (Fig. 2d) was found to
be a competitive trypsin inhibitor that lost the inhibitory activity upon incubation with trypsin. The linear peptide in which the
sulfur atoms of Cys 2 and Cys 12 were blocked has a higher inhibi-

```
   1              5                    10                      15
Phe-Thr-Phe-Thr-Tyr-Thr-Asp-Pro-Asn-Cys-Gln-Thr-Gly-Gln-Gly-Gln-Asn-

        20                    25                      30                34
Pro-Asn-Gly-Ile-Ser-Glu-Pro-Thr-Ala-Ala-Lys-Val-Gln-Ala-His-Cys-Ala
```

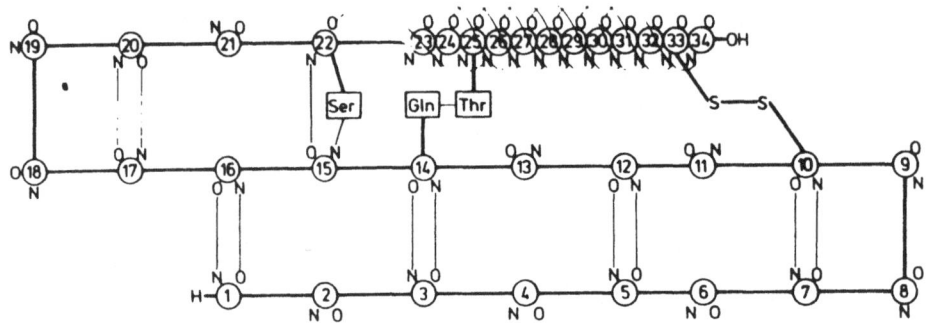

Figure 4. Amino acid sequence and proposed secondary structure of the designed synthetic 34-residue polypeptide. Thin lines represent hydrogen bonds between the β-strands and in the α-helix, amino acid symbols in boxes are side chains involved in hydrogen bonds. Reprinted by permission from Nature, Vol. 281, N° 5733, pp. 650-655. Copyright (c) 1979 Macmillan Journals Limited.

tion constant and was deactivated by trypsin slightly faster.

DESIGN AND SYNTHESIS OF POLYPEPTIDES POSSESSING ENZYMATIC ACTIVITY

Artificial enzymes are easily obtained in random processes. Thus, thermal copolymerization of amino acids yields a mixture of proteinoids differing in composition and size and capable of catalyzing chemical reactions such as hydrolyses, decarboxylations, or aminations (Rohlfing and Fox, 1969). Other examples are the synthetic copolymers of glutamic acid and hydrophobic amino acids that were shown to have substantial lysozyme-like activity (Naithani and Dhar, 1967). However, the isolation from these mixtures (Rohlfing and Fox, 1969 ; Naithani and Dhar, 1967) of single spe-

cies possessing defined enzymatic properties seems extremely difficult and has not yet been reported.

Reports in the literature on designed synthetic enzymes are
scarce. However, the results obtained so far (Gutte et al., 1979 ;
Chakravarty et al., 1973 ; Moser et al., 1983) are very encouraging
and should stimulate the interest in this new direction of protein
research.

A Decapeptide with Glycosidase Activity

In 1973, Dhar and coworkers (Chakravarty et al., 1973) described the synthesis of a decapeptide (Fig. 3) with glycosidase
activity.

1   6   12

H-Met-Thr-Phe-Ile-Arg-Pro-Asn-Val-Gly-Ala-Met-Ser-

13   18   24

Asn-Phe-Tyr-His-Tyr-Pro-Asn-Ile-Ile-Ile-Thr-Phe-OH

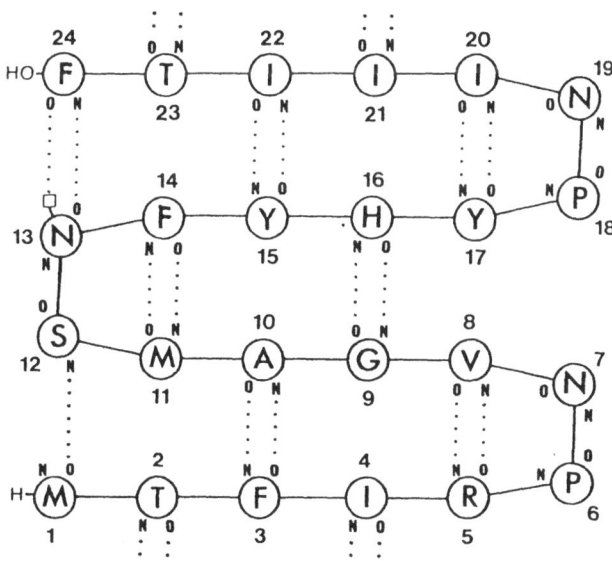

Figure 5. Amino acid sequence and proposed secondary structure of
the designed synthetic 24-residue DDT-binding polypeptide. Dotted
lines indicate hydrogen bonds between NH and CO groups.
Reprinted by permission from FEBS Lett., Vol. 157, N° 2, pp. 247-
251. Copyright (c) 1983 Federation of European Biochemical Societies.

The authors wanted to show that a model active site of lysozyme can be designed based on the known mechanism of action of the natural enzyme (Phillips, 1966) and the present-day knowledge of the rules of protein folding (Kotelchuck and Scheraga, 1969). If the synthetic decapeptide (Fig. 3) was capable of forming an α-helix, the γ'-carboxyl group of Glu 6 would be flanked by the hydrophobic side chains of Phe 2 and Phe 10. In this environment, it might be equivalent to the essential unionized carboxyl group of Glu 35 of lysozyme. On the other hand, the side chains of Glu 1 and Glu 5 would be in hydrophilic surroundings and could provide the essential ionized carboxyl group that, in the natural enzyme, is contributed by Asp 52 (Phillips, 1966). However, as a decapeptide is not expected to form a stable α-helix, enzymatic activity based on an α-helical arrangement of the active-site residues would have to be low. It is therefore not surprising that the designed decapeptide had only weak glycosidase activity using chitin and dextran as substrates.

Whether this activity was an inherent property of the decapeptide sequence chosen (Fig. 3) or whether analogues possessing the same amino acid residues in random sequences were also enzymatically active, was not tested by the authors.

Design of Proteins with Novel Enzymatic Activities

Protein design is taken a step further if the attempt is made to construct an artificial protein with a novel activity or function (e.g. a restriction endonuclease with a novel substrate specificity). Ideally, one would like to be able to synthesize tailor-made proteins for laboratory, medical, and industrial use. At present, our limited understanding of the relationship between amino acid sequence and folding of a protein and the difficulty of obtaining large synthetic polypeptide or DNA molecules are severe restrictions on the vigorous pursue of this idea. Nevertheless, we have succeeded in designing a novel 34-residue ribonuclease (Gutte et al., 1979) and a 24-residue DDT*-binding polypeptide (Moser et al., 1983) that, in the presence of heme, became a "pseudoenzyme". The basic tools for the construction of these two artificial polypeptides were a) model building and b) the secondary structure prediction rules of Chou and Fasman (1978) and Levitt (1978). In order to design an ideal α-helix or β-sheet it may be of advantage to consider additional predictive schemes (Lim, 1974 ; Garnier et al., 1978). However, as several of the residues that were incorporated for specific interactions with the ligands inevitably perturbed the ideal structures, it seemed unnecessary to search for the amino acid sequence that might produce the most stable α-helix or β-sheet. Using CPK models, the backbone of a peptide chain was folded around the ligand to determine the structural requirements

*DDT, 2,2-bis(p-chlorophenyl)-1,1,1-trichloroethane.

of a conformationally stable binding site (minimum size of the site, the kind and arrangement of secondary structural elements). The next step was choosing an amino acid sequence that, based on the prediction methods of Chou and Fasman (1978) and Levitt (1978), could produce the proposed folding and specific non-covalent bonds with the ligand. The designed polypeptides were synthesized by the solid phase method (Merrifield, 1963 ; Barany and Merrifield; Arnheiter et al., 1981).

A 34-residue polypeptide possessing ribonuclease activity

The 34-residue polypeptide was designed to bind the anticodon of yeast tRNA^{Phe} (m$_2^1$GAA) (Gutte et al., 1979). Its secondary structure and amino acid sequence (Fig. 4) were derived following the procedure described in the preceding paragraph. The main contribution to the binding of the ligand could come from stacking or intercalation interactions of the side chains of Phe 1, Phe 3, and Tyr 5 with the bases of the trinucleotide. This assumption seemed to be confirmed by circular dichroism (CD) measurements of polypeptide-2'-CMP mixtures (Jaenicke et al., 1980) and binding studies using a non-aromatic analogue of the 34-residue polypeptide and various nucleotides as ligands (Gutte, unpublished results). The near UV CD spectrum of the designed peptide was changed depending on the concentration of 2'-CMP. Replacement of the aromatic residues in positions 1, 3, and 5 with Val, Val and Ile, respectively, abolished binding of 2'-CMP and GAA almost completely.

The designed 34-residue polypeptide and its dimer containing two interchain disulfide bonds were found to interact with the trinucleotide GAA. However, the dissociation constants of complexes formed between the two peptides and cytidine phosphates were approximately two orders of magnitude lower (K_D, 10^{-4} M to 5×10^{-6} M) than those of the peptide-GAA complexes. Both peptides preferably bound cytidine phosphates ; interaction with single-stranded DNA was also strong. The difference between the binding specificity expected and that found was probably caused by deviations of the three-dimensional structure of the designed polypeptide from the folding that had been proposed (Fig. 4).

Both the 34-residue polypeptide and its covalent dimer had weak ribonuclease activity (0.1 % and 2.5 %, respectively, of the activity of bovine pancreatic ribonuclease A using tRNA as substrate) with high preference for cleavage at the 3'-end of C. The enzymatic activity of the two peptides was not anticipated. Inspection of the space-filling model of the 34-residue polypeptide showed that the side chains of three potential active site residues (His 32, Thr 12, and Asp 7) were in spatial proximity (Fig. 4). The artificial 34-residue nuclease would be a group I enzyme as defined in the introduction.

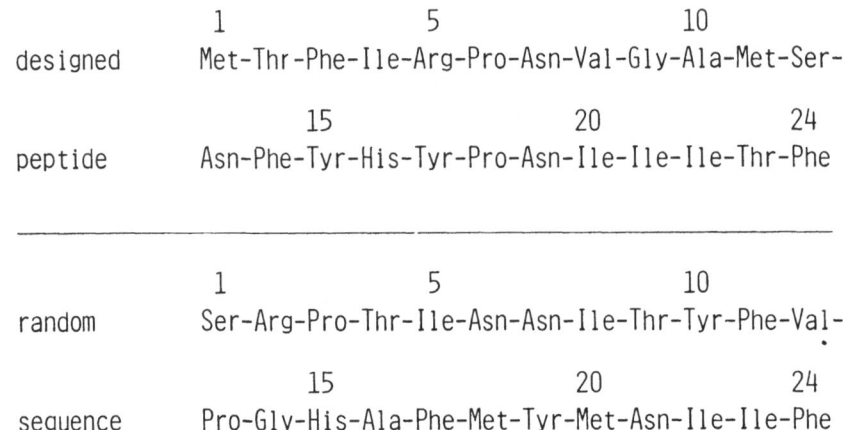

```
              1               5                    10
designed    Met-Thr-Phe-Ile-Arg-Pro-Asn-Val-Gly-Ala-Met-Ser-

              15              20                   24
peptide     Asn-Phe-Tyr-His-Tyr-Pro-Asn-Ile-Ile-Ile-Thr-Phe


              1               5                    10
random      Ser-Arg-Pro-Thr-Ile-Asn-Asn-Ile-Thr-Tyr-Phe-Val-

              15              20                   24
sequence    Pro-Gly-His-Ala-Phe-Met-Tyr-Met-Asn-Ile-Ile-Phe
```

Figure 6. Comparison of the sequences of the designed 24-residue DDT-binding polypeptide and the 24-residue analogue containing the same amino acid residues in a random order.

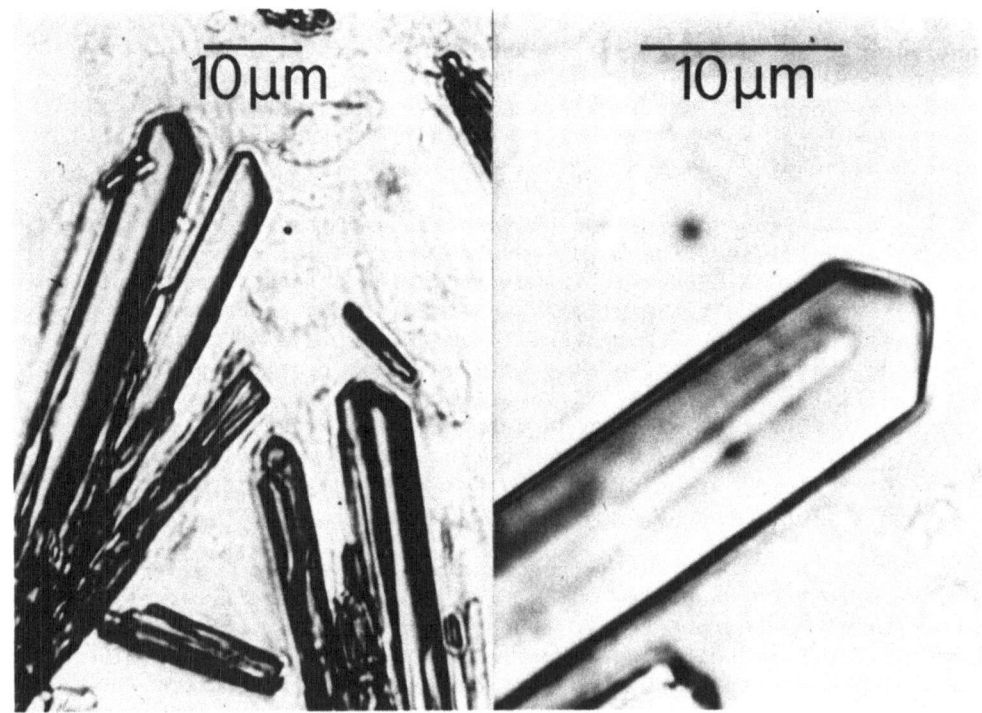

Figure 7. Crystals of the designed 24-residue DDT-binding polypeptide. Reprinted by permission from FEBS Lett., Vol. 157, N° 2, pp. 247-251. Copyright (c) 1983 Federation of European Biochemical Societies.

It is interesting to note that the dimer bound 2'-CMP 30 times more strongly than did the monomer. Likewise, the ribonuclease
activity of the dimer was 25 times higher than that of the monomer
using tRNA as substrate. Dimerization of the monomer through formation of two intermolecular disulfide bonds seemed to produce a
pocket which gave better protection of polypeptide-ligand complexes
against dissociation in aqueous solvents and in which the orientation of the substrate was more favourable for catalysis than in the
complex with the monomer.

The proposed secondary structure of the designed 34-residue
polypeptide (Fig. 4) was partly confirmed by CD measurements (Jaenicke et al., 1980, and unpublished results) yet rigorous proof of
the suggested folding and elucidation of the structure of the substrate binding site can only come from X-ray studies on suitable
crystals. So far only crystals of poor quality were obtained.

Design of a 24-residue DDT-binding polypeptide (Moser et al.,
1983) and modification of DDT by mixtures of the designed polypeptide with heme

Encouraged by the results of our first attempt to construct
a novel protein (Gutte et al., 1979), we tried to design a polypeptide with DDT-binding activity. The sequence of the residues that
could form a binding site was determined by the structure and the
hydrophobic nature of DDT and by the chain folding that had been
chosen. If DDT was bound by the designed polypeptide, possible ways
of degradation or modification of the bound ligand were to be investigated in a second stage of the work.

The proposed secondary structure (a 24-residue β-pleated
sheet consisting of four antiparallel strands) and the sequence
of the designed polypeptide are shown in Fig. 5. In the model,
a DDT binding site of high complementarity was formed by the side
chains of Ile 21, His 16, Phe 14, Met 11, and Ile 4 located on one
side of the β-sheet. The binding energy of the complex could be
provided by the stacking interactions of the aromatic rings and by
van der Waals contacts.

As DDT is insoluble in aqueous buffers, mixtures of aqueous
buffers and ethanol were used to study the DDT binding by the designed synthetic 24-residue polypeptide. In ethanol-0.05 M NH_4HCO_3
(6:5, v/v) a complex with an apparent dissociation constant of
$\approx 2 \times 10^{-5}$ M was formed (Moser et al., 1983). There was no interaction between DDT and bovine serum albumin under identical conditions. The specificity of DDT binding by the designed 24-residue
polypeptide was tested further by preparation of an analogue that
contained the same amino acid residues in a random sequence. The
primary structures of the two peptides are compared in Fig. 6.

The analogue bound DDT ≃ 100 times less strongly than did the designed peptide. Aromatic residues in alternating positions (His 15,
Phe 17, and Tyr 19) may have again contributed to the binding of
DDT. Antibodies prepared against the designed 24-residue polypeptide also reacted weakly with the 24-residue analogue. This indicated that the two peptides had structural similarity to some extent.

The designed 24-residue polypeptide was shown by preliminary
Raman spectroscopic studies to consist largely of β -structure.
Feldmann (1983) was able to model the DDT-binding polypeptide by
fitting it to four strands of the main β -sheet of concanavalin A.

The designed peptide could be crystallized (Fig. 7). Most
crystals were regular thick needles. The crystallizability indicated the high degree of purity of the synthetic product. If larger
crystals can be grown, the X-ray structure of the artificial 24-
residue DDT-binding polypeptide could be determined.

Finally our attempts to convert the DDT-binding polypeptide
into an enzyme are described. Cytochrome P-450 enzymes are known
to catalyze the hydroxylation of aromatic hydrocarbons. This is
the first of a series of reactions that convert water-insoluble
aromatic compounds like DDT into water-soluble, secretable derivatives. The prosthetic group of these enzymes is heme, one of the
ligands of the heme iron being a thiolate. Like other prosthetic
groups, heme or heme derivatives alone already have weak biological activities. A synthetic heme-imidazole compound reversibly
bound oxygen (Chang and Traylor, 1973) and both a hemin-cysteine
(Sakurai, 1980) and a hemin-glutathione complex (Sakurai et al.,
1981) were cytochrome P-450 models catalyzing the hydroxylation
of aniline (Sakurai, 1980 ; Sakurai et al., 1981), the dealkyla-
tion of anisidine and phenetidine (Sakurai et al., 1981), and the
aromatic methyl migration of p-toluidine (Sakurai et al., 1981).
The hemin-cysteine system was found to catalyze also the deriva-
tization of DDT. All products formed were more hydrophilic than
DDT as shown by thin layer chromatography of the reaction mixture.
Identification of the reaction products is under way. The expected
hydroxylation may proceed following the scheme given in Fig. 8.
In the presence of the designed 24-residue DDT-binding polypeptide
DDT modification by the hemin-cysteine system was ≃ 100 times
faster. Both the hemin-cysteine complex and DDT were preincubated
separately with the designed peptide. Then the two solutions were
mixed and the reaction was followed spectrophotometrically at 237 nm
and by thin layer chromatography. The products of the reaction were
partly different from those formed in the absence of the designed
peptide.

Perhaps we are witnessing the first steps of the evolution
of an enzyme. The combination of heme with the DDT-binding poly-

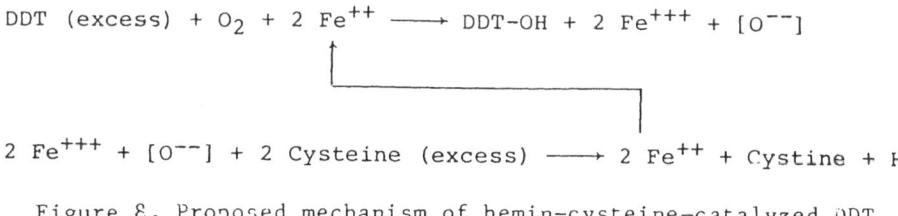

DDT (excess) + O_2 + 2 Fe^{++} ⟶ DDT-OH + 2 Fe^{+++} + [O^{--}]

2 Fe^{+++} + [O^{--}] + 2 Cysteine (excess) ⟶ 2 Fe^{++} + Cystine + H_2O

Figure 8. Proposed mechanism of hemin-cysteine-catalyzed DDT hydroxylation.

peptide led to a considerable increase of the rate of DDT modification. DDT-binding polypeptide and heme were found to form a weak complex ($K_n \simeq 10^{-3}$ M) which could be called a group II enzyme as defined in the introduction. The rate increase of the heme-catalyzed reaction in the presence of the designed synthetic peptide could have been caused by a peptide-mediated orientation of DDT and heme that was favourable for catalysis.

ACKNOWLEDGEMENTS

We thank Dr. H. Vogel (Max-Planck Institut für Biologie, Tübingen, Germany) for recording the Raman spectra. The work performed in our laboratory was supported in part by a grant to B.G. from the Schweizerische Nationalfonds.

REFERENCES

Arnheiter, H., Thomas, R.M., Leist, T., Fountoulakis, M. and Gutte, B. (1981) Nature 294, 278-280.
Barany, G. and Merrifield, R.B. (1979) in "The Peptides", Vol. 2, E. Gross and J. Meienhofer, eds., Academic Press, New York.
Blow, D.M. and Steitz, T.A. (1970) Annu. Rev. Biochem. 39, 63-100.
Chakravarty, P.K., Mathur, K.B. and Dhar, M.M. (1973) Experientia 29, 786-788.
Chang, C.K., and Traylor, T.G. (1973) Proc. Natl. Acad. Sci. USA 70, 2647-2650.
Chou, P.Y. and Fasman, G.D. (1978) Adv. Enzymol. 47, 45-148.
Feldmann, R.J. (1983) in "Computer Applications in Chemistry", S. R. Heller and R. Potenzone, Jr., eds., Elsevier, Amsterdam.
Fukushima, D., Kupferberg, J.P., Yokoyama, S., Kroon, D.J., Kaiser, E.T. and Kézdy, F.J. (1979) J. Amer. Chem. Soc. 101, 3703-3704.
Garnier, J., Osguthorpe, D.J. and Robson, B. (1978) J. Mol. Biol. 120, 97-120.
Gutte, B., Däumigen, M. and Wittschieber, E. (1979) Nature 281, 650-655.

Huber, R. and Bode, W. (1978) Accounts Chem. Res. 11, 114–122.
Huber, R., Kukla, D., Steigemann, W., Deisenhofer, J. and Jones, A.
(1974) in "Proteinase Inhibitors", H. Fritz, H. Tschesche, L. J.
Greene and E. Truscheit, eds., Springer–Verlag, Berlin.
Iyer, K.S., Laussac, J.P., Lau,S.J. and Sarkar, B. (1981) Int.
J. Peptide Protein Res. 17, 549–559.
Jaenicke, R., Gutte, B., Glatter, U., Strassburger, W. and Wollmer,
A. (1980) FEBS Lett. 114, 161–164.
Kitchell, J.P.and Dyckes, D.F. (1982) Biochim. Biophys. Acta 701,
149–152.
Komoriya, A. and Chaiken, I.M. (1982) J. Biol. Chem. 257, 2599–
2604.
Kotelchuck, D. and Scheraga, H.A. (1969) Proc. Natl. Acad. Sci.
USA 62, 14–21.
Lee, C.L. and Atassi, M.Z. (1976) Biochem. J. 159, 89–93.
Levitt, M. (1978) Biochemistry 17, 4277–4285.
Lim, V.I. (1974) J. Mol. Biol. 88, 873–894.
Means, G.E. and Feeney, R.E. (1971) "Chemical Modification of Pro-
teins", Holden–Day, San Francisco.
Merrifield, R.B. (1963) J. Amer. Chem. Soc. 85, 2149–2154.
Moser, R., Thomas, R.M. and Gutte, B. (1983) FEBS Lett. 157, 247–251.
Naithani, V.K. and Dhar, M.M. (1967) Biochem. Biophys. Res. Comm.
29, 368–372.
Phillips, D.C. (1966) Sci. Amer. 215, 78–90.
Richards, F.M. and Wyckoff, H.W. (1971) in "The Enzymes", 3rd ed.,
Vol. IV, P.D. Boyer, ed., Academic Press, New York.
Rohlfing, D.L. and Fox, S.W. (1969) Adv. Catalysis 20, 373–418.
Sakurai, H. (1980) Chem. Pharm. Bull. 28, 3437–3439.
Sakurai, H., Shimomura, S. and Ishizu, K. (1981) Biochem. Biophys.
Res. Comm. 101, 1102–1108.
Veber, D.F., Freidinger, R.M., Perlow, D.S., Paleveda Jr., W.J.,
Holly, F.W., Strachan, R.G., Nutt, R.F., Arison, B.H., Homnick, C.,
Randall, W.C., Glitzer, M.S., Saperstein, R. and Hirschmann, R.
(1981) Nature 292, 55–58.

KINETICS OF COMPLEX SELFREPLICATING MOLECULAR SYSTEMS

Peter Schuster

Institut für Theoretische Chemie und Strahlenchemie
Universität Wien, Währingerstrasse 17, A 1090 Wien

AUTOCATALYSIS AND SELFREPLICATION

Some important properties of biological systems like selection and evolutionary optimization are observed also in chemical reaction networks when they contain autocatalytic processes. Autocatalysis in almost all known cases cannot be traced down to single elementary steps. The autocatalytic effect is a net result of several individual reactions. For illustration we present two characteristic and well studied examples from the inanimate world: the vapor phase formation of water from hydrogen and oxygen and the Belousov-Zhabotinsky reaction (Figure 1). Polynucleotide replication in this respect is no exception : the most detailed studies of the mechanism were performed on RNA replication catalyzed by a virus specific replicase. The single strand RNA bacteriophage Qβ was chosen as a suitable model system (Biebricher et al, 1981, 1982 and 1983).

The mechanism which has been proposed to fit all experimental data available is shown in Figure 2. It consists of a many step polymerization process. In addition, the dynamics of RNA replication is complicated by complementary copying : the plus strand serves as a template for the synthesis of the minus strand and vice versa. Under the experimental conditions applied, double strand formation represents a "dead end" of the replication process. Double helices are not accepted as templates by the enzyme and they do not dissociate to a sufficient extent. Thus, one important role of the Qβ replicase is to avoid double strand formation.

$$2 \, H_2 + O_2 \longrightarrow 2 \, H_2O$$

$OH + H_2$	$\longrightarrow H_2O + H$	(1)
$H + O_2$	$\longrightarrow OH + O$	(2)
$O + H_2$	$\longrightarrow OH + H$	(3)
$H + O_2 + M$	$\longrightarrow HO_2 + M$	(4)

HO_2	$\xrightarrow{\text{SURFACE}} \text{DESTRUCTION}$	(5)
$HO_2 + H_2O_2$	$\longrightarrow H_2O + O_2 + OH$	(6)
$H_2O_2 + M$	$\longrightarrow OH + OH + M$	(7)
$HO_2 + H_2$	$\longrightarrow H_2O_2 + H$	(8)
$HO_2 + H_2$	$\longrightarrow H_2O + OH$	(9)

M, M' ARE COLLISION PARTNERS REMOVING
EXCESS ENERGY

$$2H^+ + Br^- + BrO_3^- \rightleftharpoons HOBr + HBrO_2 \tag{1}$$
$$H^+ + HBrO_2 + Br^- \rightleftharpoons 2HOBr \tag{2}$$
$$HOBr + Br^- + H^+ \rightleftharpoons Br_2 + H_2O \tag{3}$$
$$CH_2(COOH)_2 \rightleftharpoons (OH)_2C\!=\!CHCOOH \tag{4}$$
$$Br_2 + (OH)_2C - CHCOOH \rightleftharpoons H^+ + Br^- + BrCH(COOH)_2 \tag{5}$$
$$HOBr + (OH)_2C - CHCOOH \rightleftharpoons H_2O + BrCH(COOH)_2 \tag{6}$$
$$HBrO_2 + BrO_3^- + H^+ \rightleftharpoons 2BrO_2^{\cdot} + H_2O \tag{7}$$
$$BrO_2^{\cdot} + Ce^{(III)} + H^+ \rightleftharpoons Ce^{(IV)} + HBrO_2 \tag{8}$$
$$Ce^{(IV)} + BrO_2^{\cdot} + H_2O \rightleftharpoons BrO_3^- + 2H^+ + Ce^{(III)} \tag{9}$$
$$2HBrO_2 \rightleftharpoons HOBr + BrO_3^- + H^+ \tag{10}$$
$$Ce^{(IV)} + CH_2(COOH)_2 \longrightarrow {\cdot}CH(COOH)_2 + Ce^{(III)} + H^+ \tag{11}$$
$${\cdot}CH(COOH)_2 + BrCH(COOH)_2 + H_2O \longrightarrow$$
$$Br^- + CH_2(COOH)_2 + HO\overset{\cdot}{C}(COOH)_2 + H^+ \tag{12}$$
$$Ce^{(IV)} + BrCH(COOH)_2 + H_2O \longrightarrow$$
$$Br^- + HO\overset{\cdot}{C}(COOH)_2 + Ce^{(III)} + 2H^+ \tag{13}$$
$$2HO\overset{\cdot}{C}(COOH)_2 \longrightarrow HOCH(COOH)_2 + O\!=\!CHCOOH + CO_2 \tag{14}$$
$$Ce^{(IV)} + HOCH(COOH)_2 \longrightarrow HO\overset{\cdot}{C}(COOH)_2 + Ce^{(III)} + H^+ \tag{15}$$
$$Ce^{(IV)} + O\!=\!CHCOOH \longrightarrow O\!=\!\overset{\cdot}{C}COOH + Ce^{(III)} + H^+ \tag{16}$$
$$2O\!=\!\overset{\cdot}{C}COOH + H_2O \longrightarrow O\!=\!CHCOOH + HCOOH + CO_2 \tag{17}$$
$$Br_2 + HCOOH \longrightarrow 2Br^- + CO_2 + 2H^+ \tag{18}$$
$$HOBr + HCOOH \longrightarrow Br^- + H^+ + CO_2 + H_2O \tag{19}$$
$$2 \cdot CH(COOH)_2 + H_2O \longrightarrow CH_2(COOH)_2 + HOCH(COOH)_2 \tag{20}$$
$$Br_2 + BrCH(COOH)_2 \longrightarrow Br_2CHCOOH + Br^- + H^+ + CO_2 \tag{21}$$
$$HO\overset{\cdot}{C}(COOH)_2 + BrCH(COOH)_2 + H_2O \longrightarrow$$
$$Br^- + HOCH(COOH)_2 + HO\overset{\cdot}{C}(COOH)_2 + H^+ \tag{22}$$

Fig. 1. Many step mechanisms of two autocatalytic reactions in reality : vapor-phase oxidation of molecular hydrogen (a) and the Belousov-Zhabotinski reaction in aqueous solution (b).

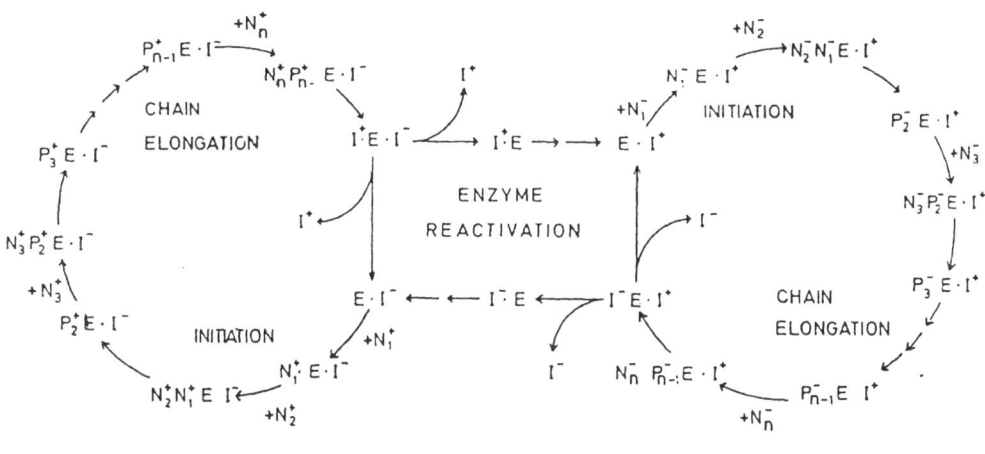

$$I^+ \cdot E \rightleftarrows I^+ + E \rightleftarrows E \cdot I^+$$

$$I^- \cdot E \rightleftarrows I^- + E \rightleftarrows E \cdot I^-$$

Fig.2. A cyclic mechanism of bacteriophage RNA replication in vitro as used by Biebricher et al (1982, 1983) to interpret the kinetic results as obtained with the Qβ system. E, I^+ and I^- represent free Qβ-polymerase, and the plus and minus strand of the RNA to be replicated. N_1^+, N_2^+, ..., N_n^+ and N_1^-, N_2^-, ..., N_n^- are the nucleoside triphosphates in a sequence as they appear in the plus and minus strand respectively. The chain of the newly synthesized molecule grows always from the 5'- to 3'- end. P_2^+, P_3^+,..., P_{n-1}^+ and P_2^-, P_3^-,..., P_{n-1}^- are used as symbols for the growing chains. A nucleation length of two bases is assumed. Enzyme reactivation is necessary because the two polynucleotides are not in suitable positions to restart polymerization : the newly synthesized strand is bound with the correct end (3') but in the wrong site (synthetic site) whereas the old strand is sitting in the correct (reading) site but bound with the wrong end (5'). The many step mechanism thus consists of three distinguishable processes : initiation of RNA synthesis, chain propagation and reactivation of the enzyme.

The time course of polynucleotide concentration in a typical replication experiment is shown schematically in figure 3. Under large excess of monomers GTP, ATP, CTP and UTP one can distinguish three phases characterized by different growth dynamics of RNA concentration :

(1) At low polynucleotide concentration enzyme molecules are in excess $(C_E^0 > C_N^0)$ and every newly synthesized polynucleotide binds instantaneously to a replicase molecule and, thus, starts the replication cycle. As a consequence of enzyme excess, the growth behaviour is determined by the RNA concentration which increases exponentially.

(2) At polynucleotide concentrations larger than total enzyme concentration $(C_E^0 < C_N^0)$ every enzyme molecule is bound to an RNA template. The rate of RNA synthesis is determined now by the concentration of the replicase and the RNA concentration grows linearly.

(3) At very large polynucleotide concentrations $(C_E^0 \ll C_N^0)$ we observe product inhibition of the enzyme and the rate of increase of the RNA concentration is slower than linear.

In order to explain the three phases of growth of RNA concentration it is not necessary to consider the polymerization process in detail. The time course of the concentration shown in Figure 3 can be described also by a simple mechanism which assumes quasiequilibrium for the individual chain elongation steps (Gassner and Schuster, 1982). In this case we distinguish only six different

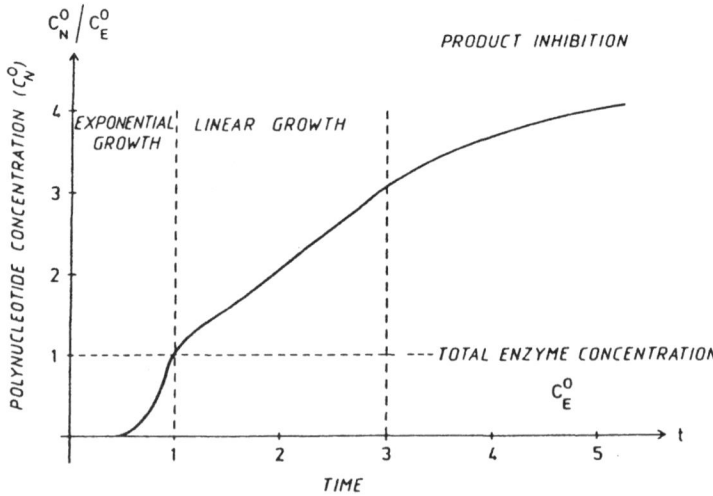

Fig.3. RNA synthesis in the test tube. A characteristic experimentally recorded curve shows three phases of growth: exponential growth at low polynucleotide concentration, linear growth above saturation of the enzyme by polynucleotides and further levelling off at still higher concentrations of template when enzyme reactivation becomes the rate determining step.

complexes between polynucleotide and replicase molecules (Figure 4). Only few details of the curve $C_N^0(t)$ are not explainable without explicit consideration of individual steps of the polymerization mechanism. One of these features is the kink in the curve at the point $C_N^0 = C_E^0$. In the simplified mechanism the tangent of the exponential curve at this point coincides with the linear continuation to higher polynucleotide concentrations.

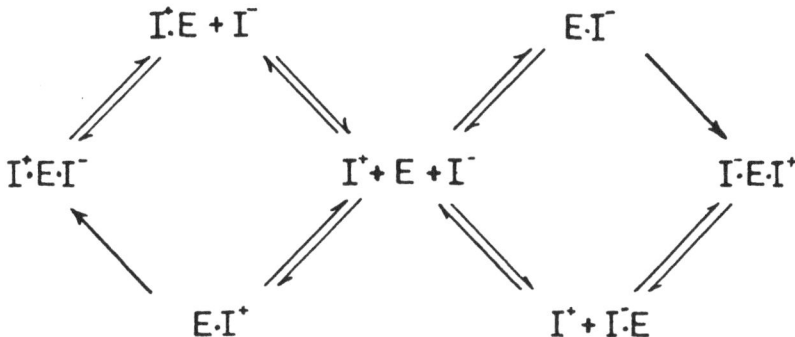

Fig.4. A simplified cyclic mechanism of bacteriophage RNA replication derived from figure 2 by neglect of the individual chain elongation steps. All reactions with the exception of the polymerization process are assumed to occur reversibly. The simplified mechanism is capable of reproducing the whole curve shown in Figure 3 except the kink at the point of enzyme saturation $(C_N^0 = C_E^0)$. As in Figure 2 we denote the plus- and minus- strand of the RNA by I^+ and I^- and the free Qβ-polymerase by E. Six different complexes of the three macromolecules are formed : I^+, $E.I^-$, $E.I^+$, $E.I^-$, $I^+.E.I^-$ and $I^-.E.I^+$. They differ with respect to the binding sites on the RNA (3' or 5' end) and on the enzyme (synthetic or reading site).

EXPERIMENTAL DATA ON SELECTION IN THE Qβ SYSTEM

The first studies on selection in an ensemble of RNA molecules in the test tube were performed by Spiegelman and coworkers (for a review of the early work see Spiegelman, 1971). RNA of the simple bacteriophage Qβ was transferred into a medium which is most favourable for RNA synthesis. This medium, mentioned already in the preceding section, consists of a solution of nucleoside triphospha-

tes and of the specific polymerase of Qβ, both in excess. In this
solution RNA synthesis starts instantaneously after template has
been added.

In the test tube experiments an open system is created by the
serial transfer technique (Figure 5) : after a certain period of
replication a small sample containing RNA from one test tube is
transferred into the next one containing fresh solution. Thereby,
the material consumed by the replication process is renewed. The
whole procedure is repeated many times. The most interesting
result of these experiments is a spontaneous increase in the
rate of RNA synthesis which occurs several times before an
optimum rate is attained. The increase in the replication rate is
accompanied by a decrease in the molecular weight of the RNA.

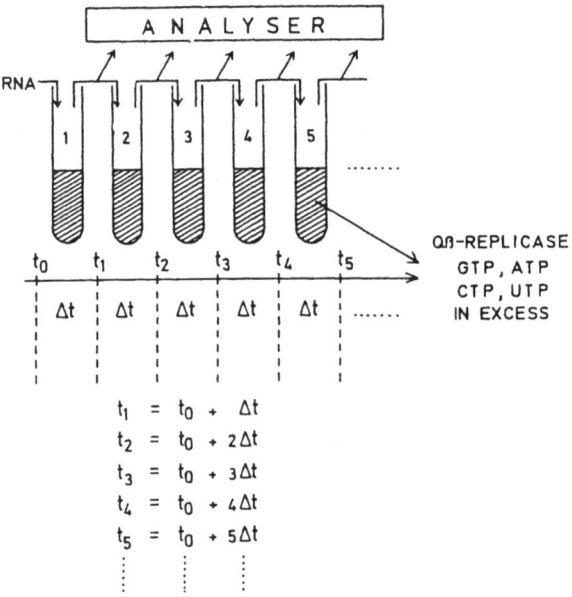

Fig. 5. The technique of serial transfer experiments (Spiegelman,
1971). RNA, in particular RNA of the bacteriophage Qβ, grows in a
medium which contains the enzyme Qβ replicase as well as the four
nucleoside triphosphates ATP, UTP, GTP and CTP in excess. After a
time Δt a sample is taken out in the test tube . Part of it is
analyzed, part of it is transferred into a new medium. The procedure is repeated after time intervals of Δt. The conditions are
chosen such that the RNA is growing in the exponential phase (see
Figure 3).

More recently, it was shown in the laboratory of Eigen
(Sumper and Luce, 1975; Biebricher et al, 1981) that an analogous
process can be carried out from lower to higher molecular weights

of the RNA as well. Starting from highly purified Qβ replicase,
i.e. an enzyme sample without any detectable impurity of polynucleotides[*] , and an excess of triphosphates GTP, ATP, CTP and UTP,
RNA is synthesized de novo . In serial transfer experiments these
de novo products show increasing rates of RNA synthesis . Thereby,
the increase in rate is accompanied by an increase in molecular
weight. The optimum rates of RNA synthesis and the molecular weights
ultimately attained in these experiments are very close to those
of serial transfer experiments starting from high molecular weight
RNA samples.

   RNA replication under the conditions of serial transfer experiments, thus, has a defined optimum rate which depends on temperature, ionic strength and other experimental parameters. What one
observes is optimization of the rate of RNA replication in the
sense of Darwin's theory. Accidental replication errors leading
to RNA which replicates faster are amplified through selection.
Most replication errors will lead to less efficient RNA molecules
and, hence, are instantly discarded. For more details on test tube
evolution the interested reader is referred to the recent review by
Biebricher (1983).

PREREQUISITES OF EVOLUTIONARY OPTIMIZATION

   Evolutionary optimization of replication rates is achieved
through selection of more efficiently replicating molecules. In
order to study the conditions under which selection occurs we
conceive a model system which is simple enough to be investigated
by analytical techniques. The autocatalytic processes are represented by single elementary steps (1). In addition, we assume degradation reactions (2) and a recycling process (3) which is dependent
on an external energy source and converts the degradation products
back into energy rich monomers (figure 6)

$$(1) \qquad A + I_k \underset{f'_k}{\overset{f_k}{\rightleftharpoons}} 2I_k \; ; \quad k=1,2,\ldots,n$$

---

[*]A recent criticism of this de novo RNA synthesis with Qβ-replicase
by Hill and Blumenthal (1983) is off the point since the experiments
which were used to disprove the onset of RNA synthesis in highly
purified enzyme fractions were carried out under conditions under .
which de novo synthesis was not detectable within the duration of the
experiment (Biebricher et al, 1981).

(2)
$$I_k \; \overset{d_k}{\underset{d_k'}{\rightleftharpoons}} \; B \quad ; \; k=1,2,\ldots,n$$

(3)
$$B \; \overset{g(E)}{\longrightarrow} \; A$$

For our purpose here, it is sufficient to consider a single substance A representing the energy rich material and a single degradation product B. By I_k, k=1, 2,...,n we denote different polynucleotides which compete for the same energy source (A). The recycling
reaction (3) is dependent on an external energy source (E) and hence
can be controlled from the outside. Reaction (3) drives the system
off equilibrium and, hence, g(E) can be understood as a measure of
the distance from thermodynamic equilibrium in this open system.

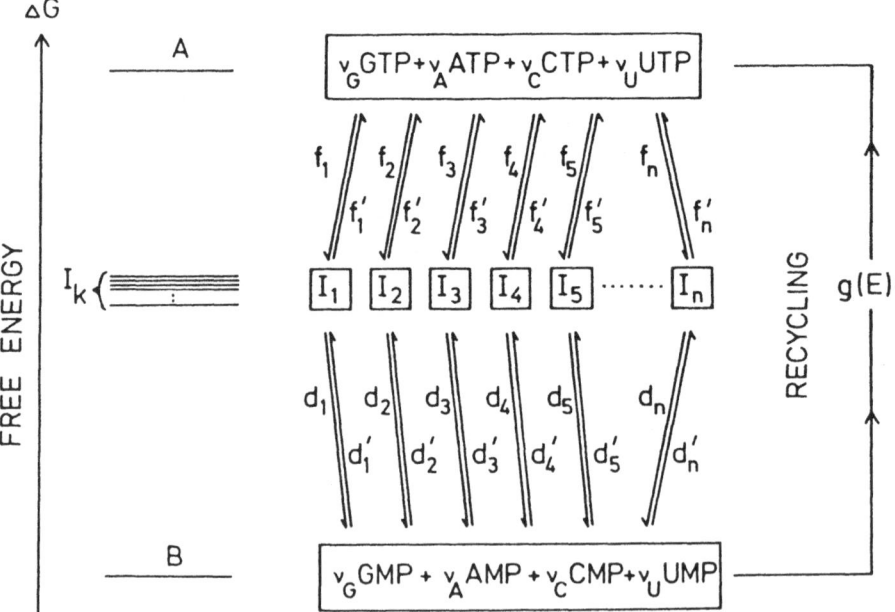

Fig.6. The thermodynamics of polynucleotide synthesis and degradation. In order to observe selection both reactions have to be
accompanied by large negative changes in Gibbs free energy ($\Delta G < 0$).
For the details of the reaction mechanisms see sections II and III.

The mechanism (1) to (3) represents a chemical reaction network of 2n+1 individual reaction steps. It can be represented by the
differential equations (4) to (6) :

(4) $\dot{a} = gb + \sum_k x_k (f'_k x_k - f_k a)$

(5) $\dot{x}_k = x_k (f_k a - f'_k x_k - d_k) + d'_k b$

(6) $\dot{b} = \sum_k d_k x_k - (g + \sum_k d'_k) b$

with k=1,2,...,n. Herein, we denoted concentrations by small letters:
$[A] = a$, $[B] = n$ and $[I_k] = x_k$. We have a conservation law

(7) $a + b + \sum_k x_k = c_o = cont.$

thus leaving us with n+1 degrees of freedom. It is important to
obtain results which are valid for arbitrary n. Therefore, we performed complete qualitative analysis on equations (4) to (6).
Since the details have been presented previously (Hofbauer and
Schuster, 1984) we restrict ourselves here to a short enumeration
of the most important results.

The dynamical system (4)-(6) depends on two external parameters g and c_o. Within the physically meaningful concentration
space* , it has two stationary states. At the state P all autocatalysts I_k are present in non-zero concentrations. This state conver-
ges to the thermodynamic equilibrium in the limit g → 0. The
second state, the "zero state" P_o is characterized by vanishing
concentrations of all autocatalysts I_k and of B ($\bar{a} = c_o$). Stability
in this dynamical system is a global property : only one of the
two states is asymptotically stable for a given pair of values of
the external parameters g and c_o (Figure 7). This scenario corresponds to an autocatalytic process below the threshold of ignition.
Above a certain value of c_o that state P is stable independently
of the value g.

---

*The physically meaningful domain of concentration variables is
defined : $0 \leqslant a \leqslant c_o$; $0 \leqslant b \leqslant c_o$; $0 \leqslant x_k \leqslant c_o$, k=1,2,...,n with
$a+b+\Sigma x_k = c_o$. Hence, the concentration space can be mapped onto a
simplex S_{n+2} .

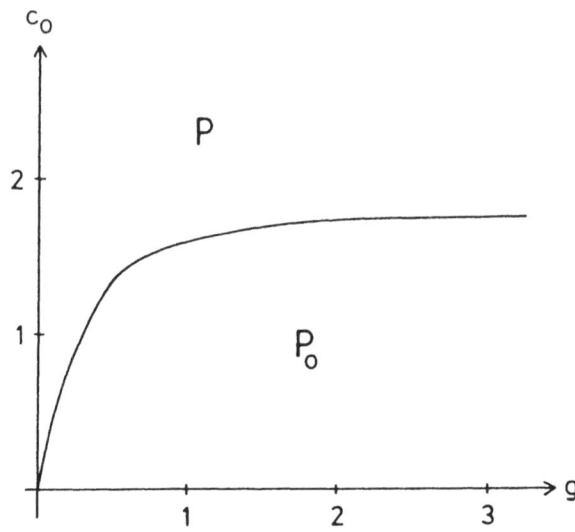

Fig. 7. Regions of asymptotic stability of different stationary
states of the differential equations (4)-(6) in the plane of external parameter (g,c_0). P_0 indicates asymptotic stability of the
"zero state", P asymptotic stability of the "thermodynamic branch".
The bifurcation resulting in an exchange of stability between P_0
and P occurs at the curve. Here we assumed reversibility of
polynucleotide synthesis and degradation ($d=d'=f=f'=1$). All rate
constants and concentrations are given in arbitrary time and
concentration units.

    In this most general form the mechanism (1) to (3) does not
lead to selection[*]. Therefore, we shall search now for special
combinations of the rate constants f_k, f_k', d_k and d_k' which eventually introduce selection into equations (4)-(6).   At first we
try "practical irreversibility" of the degradation process which
can be expressed as the limit $\lim d_k' \to 0$ for all $k=1,2,..,n$. By
the term "practical irreversibility" we indicate that the process
is run under conditions at which ΔG is negative and large in
absolute value (Figure 6). Then the reverse reaction is too slow
for experimental observation. The equilibrium constant of the
reaction, nevertheless, is still finite. The number of stationary

---

[*]In precise mathematical terms we define selection as a process
leading to a final state at which precisely one autocatalyst (I_m)
is present : $\lim\limits_{t\to\infty} x_m = c_0 - a - b$ and $\lim\limits_{t\to\infty} x_k = 0$ for all $k\neq m$.

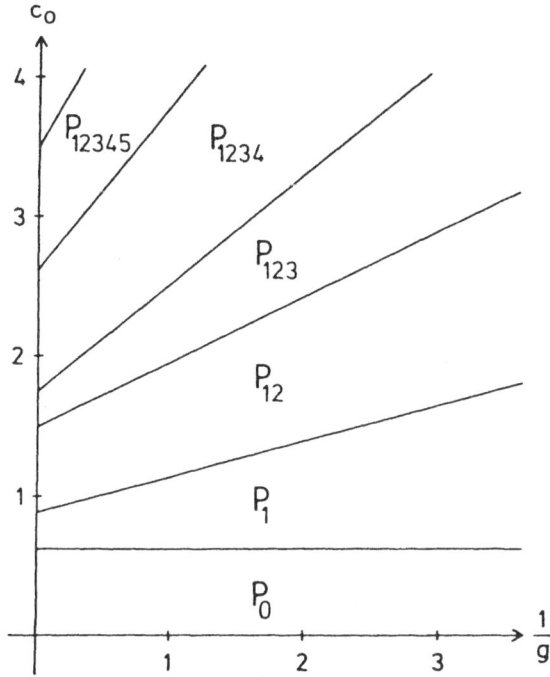

Fig.8. Regions of asymptotic stability of different stationary
states of the differential equation (4)-(6) with irreversible
degradation ($d_k' = 0$; $k=1,2,\ldots,n$) in the plane of external parameters
(g, c_o). P_o indicates asymptotic stability of the "zero state".P_1
selection of the most efficient polynucleotide I_1, P_{12} the simultaneous presence of I_1 and I_2 at the stationary state etc. Bifurcations occur at the straight lines separating regions of different
stationary states. We are dealing here with the most simple type
of bifurcations exclusively : stationary states exchange stability.

states on the concentration simplex now is much larger than in the
system with reversible degradation: we find a stationary state also
on every subspace of the autocatalyst concentrations, e.g. P_i is a
steady state at which all autocatalysts but I_i are absent, P_{ij}
is the state at which only I_i and I_j are present. The sequence of
states is continued until we reach finally the state $P_{12\ldots n}$ at
which the concentrations of all autocatalysts I_1, I_2, $\ldots$, I_n
are positive. Stable stationary states are unique as in the general
case ($d_k' > 0$): only one steady state is asymptotically stable
for a given combination of external parameters. Without losing
generality we order the competitively replicating polynucleotides

in a sequence of increasing ratio of rate constants d_k/f_k[*] :

$$\frac{d_1}{f_1} < \frac{d_2}{f_2} < \frac{d_3}{f_3} < \ldots < \frac{d_n}{f_n}$$

In this case the sequence of stable stationary states with increasing total concentration c_o is :

$$P_o, \; P_1, P_{13} \;, \; P_{23}, \; \ldots, \; P_{12\ldots n}$$

Figure 8 shows regions of asymptotic stability of these stationary states. In the region in which P is stable we observe selection of I_1. This is the polynucleotide with the smallest ratio of the rate constants for degradation and autocatalytic reproduction :

$$\frac{d_1}{f_1} = \min \; (\; \frac{d_k}{f_k}; k=1,2,\ldots,n) \to \lim_{t\to\infty} x_1 = c_o - a - b; \; \lim_{t\to\infty} x_k = 0, \; k=2,3,\ldots,n$$

This ratio of rate constants thus represents the selection criterion in the reaction network under consideration. From Figure 8 we derive also general trends concerning the numbers of coexisting polynucleotide sequences: the larger the total concentration c_o and the larger the rate constant g of the recycling reaction are , the more polynucleotides are coexistent.

Let us now search for conditions which make the region of selection as large as possible. From Figure 8 and the corresponding conditions for stability of steady states we compute the point at which the line separating the stability regions of P_1 and P_{12} intersects the c_o axis :

$$(8) \qquad \lim_{g\to\infty}(P_1 \cap P_{12}) = \Delta_1 = \frac{d_2}{f_2} + \frac{f_1}{f_1'} (\frac{d_2}{f_2} - \frac{d_1}{f_1})$$

Hence, Δ_1 approaches $+\infty$ as f_1' goes to zero. The region of selection, stability of P1, becomes infinitely large when replication occurs under conditions of practical irreversibility.

In order to avoid transients leading towards a stable zero

---

[*]We dispense here with a discussion of kinetically degenerate cases which have been studied elsewhere (Schuster and Sigmund, 1983).

state P_o, it is sufficient to introduce a large reservoir of A
which sustains a constant concentration a_o. Then selection occurs
for all c_0 values and for almost all initial conditions. The
selection criterion is slightly changed now :

$$(f_1 a_o - d_1) = max(f_k a_o - d_k; k=1,2,\ldots,n) \rightarrow \lim_{t\to\infty} x_1 \neq 0, \lim_{t\to\infty} x_k = 0, k=2,3,\ldots,n$$

In summarizing this section we present four prerequisites for the occurrence of selection in systems if replication molecules:
(1) Selection occurs exclusively in open systems. As far as the
reaction mechanism (1)-(3) is concerned this requires a non-vani-
shing recycling process : $g > 0$.
(2) Selection requires an irreversible degradation process. In
systems with reversible degradation ($d_k' > 0$) we are dealing always
with finite concentrations of all competitors at the non-trivial
stationary states.
(3) Selection occurring in a wide range of external parameters
requires an irreversible replication process. The smaller the rate
constant f_1' for the reverse replication process of the most efficient autocatalyst I_1 is, the larger is the range in the g,c_0-plane
within which I_1 is selected.
(4) Selection occurs under almost all conditions if the reservoir
of energy rich material is such large that the concentration of A
is constant for practical purposes.

The conditions (1) to (4) represent a kind of hierarchy, the
first ones being more severe than the latter ones :
(1) and (2) are conditions "sine quibus non" for selection; (3) and
(4) only enlarge the range of external parameters within which
selection occurs. All four conditions are fulfilled in a kind of
dialysis reactor shown in Figure 9 which has been postulated in
the original derivation of the selection equations (Eigen, 1971;
Eigen and Schuster , 1979; Küppers, 1979).

It is interesting to realize that conditions (2) and (3) are
indeed fulfilled by present-day cellular biochemistry. Both, the
synthesis of DNA and RNA from nucleoside triphosphates as well
as the degradation of polynucleotides into nucleotides monophosphates are accompanied by large negative ΔG values.

ADAPTATIVE SELECTION, RANDOM REPLICATION AND NEUTRAL MUTATIONS

Although template induced RNA polymerization is a complicated
many step process (Fig. 2), it can be mimicked by simple overall
reactions provided the dialysis reactor (Fig. 9) is run under such
conditions that the polymerization is in the phase of exponential
growth (Fig. 3):

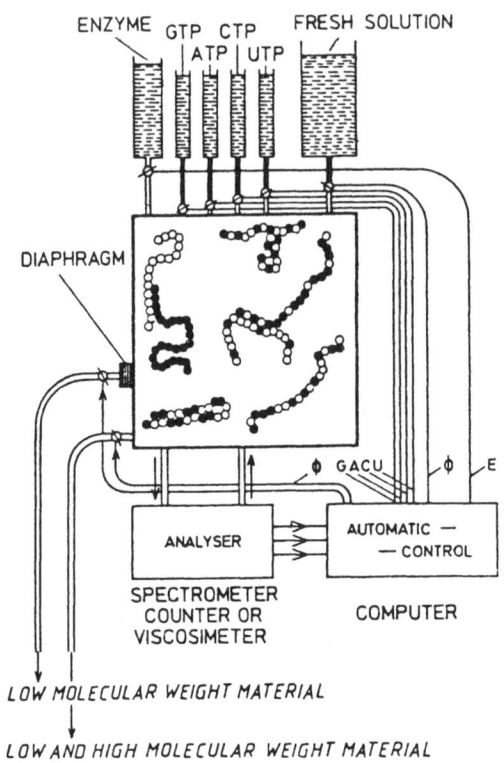

Fig. 9. A dialysis reactor for evolution experiments (see also Küppers, 1979). This kind of flow reactor consists of a reaction vessel which allows for temperature and pressure control. Its walls are impermeable to polynucleotides. Energy rich material is poured from the environment into the reactor. The degradation products are removed steadily. Material transport is adjusted in such a way that the concentration of monomers is constant in the reactor. A dilution flux ϕ is installed in order to remove the excess of polynucleotides produced by multiplication. Thus the sum of the concentrations :

$$[I_1] + [I_2] + \ldots + [I_n] = \sum_{i=1} x_i = c \text{ may be controlled by the flux } \phi.$$

Under "constant organization" ϕ is adjusted such that the concentration $c = c_0$ is constant. The regulation of ϕ requires internal control which may be achieved by analysis of the solution and data processing by a computer, as indicated above. In the examples studied here ϕ is directly related to the rate constants of polynucleotides synthesis and degradation :

$$\phi = c_0 \bar{E} = \sum_k (f_k - d_k) x_k .$$

$$(9) \qquad I_k + \sum_{\lambda=1}^{4} \nu_{k\lambda} A_\lambda \xrightarrow{\quad f_k Q_{kk} \quad} 2I_k$$

$$(10) \qquad I_j + \sum_{\lambda=1}^{4} \nu_{k\lambda} A_\lambda \xrightarrow{\quad f_j Q_{kj} \quad} I_j + I_k$$

$$(11) \qquad I_k \xrightarrow{\quad d_k \quad} \sum_{\lambda=1}^{4} \nu_{k\lambda} B_\lambda$$

$$j,k = 1,2,\ldots,n$$

By I_k we denote a particular polynucleotide sequence. We have n different polynucleotide sequences in our system. In practice n may be very large and difficult to determine experimentally. A_λ are the nucleoside triphosphates GTP, ATP, CTP and UTP, B_λ the corresponding monophosphates GMP, AMP, CMP and UMP. Stoichiometric coefficients of the gross polymerization reaction are denoted by $\nu_{k\lambda}$. In the reaction mechanism shown above we consider three classes of reactions :
(1) template induced faithful replication (9),
(2) erroneous replication (10) and
(3) hydrolytic cleavage of the polynucleotide (11).

The rate constants f_k give the gross numbers of copies, correct and erroneous, synthesized from template I_k per unit time and unit concentrations. The quality factor Q_{kk} is a measure of the accuracy of the replication process : Q_{kk} means perfect, error-free replication. The frequency at which I_k is obtained as an error copy of I_j is denoted by Q_{kj}. Accordingly, we have a conservation law

$$(12) \qquad \sum_{k=1}^{n} Q_{kj} = 1$$

since every copy has to be either correct or erroneous. By d_k, finally, we denote the rate constants of the degradation process.

In case all polynucleotides are present in sufficiently large numbers -- in formal mathematical terms this is the case in an

infinite population – the reactions (9), (10) and (11) run in the
dialysis reactor shown in Figure 9 can be described by the conventional kinetic equations :

(13) $\dot{x}_k = \dfrac{dx_k}{dt} = (w_{kk} - \bar{E})x_k + \sum\limits_{j \neq k} w_{kj}x_j$; $j,k=1,2,\ldots,n$.

Again we denote concentrations by $[I_k] = x_k$. The quantities w_{kk}
and w_{jk} are functions of rate constants, mutation frequencies
and the constant concentrations of energy rich monomers,
$[A_1] = a_1^o$, $[A_2] = a_2^o$, $[A_3] = a_3^o$ and $[A_4] = a_4^o$:

(14a) $w_{kk} = \bar{f}_k Q_{kk} - d_k$

and

(14b) $w_{kj} = \bar{f}_j Q_{kj}$

with

(14c) $\bar{f}_k = f_k (a_1^o)^{\nu_{k1}} (a_2^o)^{\nu_{k2}} (a_3^o)^{\nu_{k3}} (a_4^o)^{\nu_{k4}}$

The mean excess production $\bar{E}$ is related to the dilution discussed
in figure 9 and to the reaction rate constants

(14d) $\bar{E} = \dfrac{1}{c_o} \sum\limits_{k} (\bar{f}_k - d_k) x_k = \dfrac{1}{c_o} \phi$

The constant total concentration of polynucleotides is denoted by

(14e) $c_o = \sum\limits_{k} x_k$

The mathematical analysis of equation (13) has been presented in
great detail (Eigen, 1971; Thompson and McBride , 1974; Jones
et al 1976; Eigen and Schuster, 1979; Swetina and Schuster,1982)
and we need not repeat it here. We discuss briefly the most important results by means of three characteristic examples.

     The first example consists of an ensemble of polynucleotides
with different values of the excess production $E_k = \bar{f}_k{}' - d_k$.
Replication occurs with ultimate accuracy ($Q_{kk}=1$, $Q_{jk} = 0$ for $k \neq j$).
In this case we observe selection of the sequence I_m which is
characterized by the largest excess production:

(15) $$E_m = \bar{f}_m - d_m = \max(\bar{f}_k - d_k; \ k=1,2,\ldots,n)$$

During the selection process the mean excess production $\bar{E}$ increases monotonically. When a new mutant, say I'_m, with larger excess production, $E'_m > E_m$, appears in the reactor, I_m is replaced by this new more efficiently replicating polynucleotide I'_m. A characteristic plot of this selection process is shown in figure 10. We call it characteristicly "adaptive selection". It is representative for the phenomenon Darwin denoted "natural selection" under the particularly simple environmental conditions of the dialysis reactor.

As our second example we consider a system with non-zero mutation frequencies, $Q_{kj} > 0$. Still, we assume absence of kinetic degeneracy : all w_{kk}-values are different. Depending on the accuracy of replication we observe two different characteristic scenarios : (1) The replication process is accurate enough : the quality factor of the master sequence I_m defined by equation (15) exceeds a certain minimum value which is a function of all rate constants $\bar{f}_k$ and d_k, $k=1,\ldots,n$ (Eigen, 1971; Eigen and Schuster, 1979; Swetina and Schuster, 1982)

(16a) $$Q_{mm} > Q_{min} = \sigma_m^{-1}$$

(16b) $$\sigma_m = \frac{\bar{f}_m}{d_m + \bar{E}_{-m}} \ ; \ \bar{E}_{-m} = \sum_{i \neq m} E_i x_i / \sum_{i \neq m} x_i$$

Then we observe a kind of selection. But, instead of single polynucleotide, the sequence I_m in the previous case, a whole ensemble of sequences is present at the stable steady state. Such a stationary ensemble has been called a "quasispecies" because of some analogy to the notion of a species in biology. It consists of the master copy I_m and its most frequent mutants.
(2) The accuracy of the replication process is too low in order to sustain a stable defined stationary mutant distribution : $Q < Q_{min}$. Then, the mechanism of inheritance breaks down : all sequences are present according to their statistical weights in the stationary solution of equation (13). In case the matrix Q is symmetric ($Q_{kj} = Q_{jk}$) we find equipartition of sequences. Otherwise a mutant distribution different from equal amounts is obtained, but still it is exclusively determined by the structure of the matrix Q. The kinetic parameters ($\bar{f}_k$, d_k) have no influence on the stationary mutant

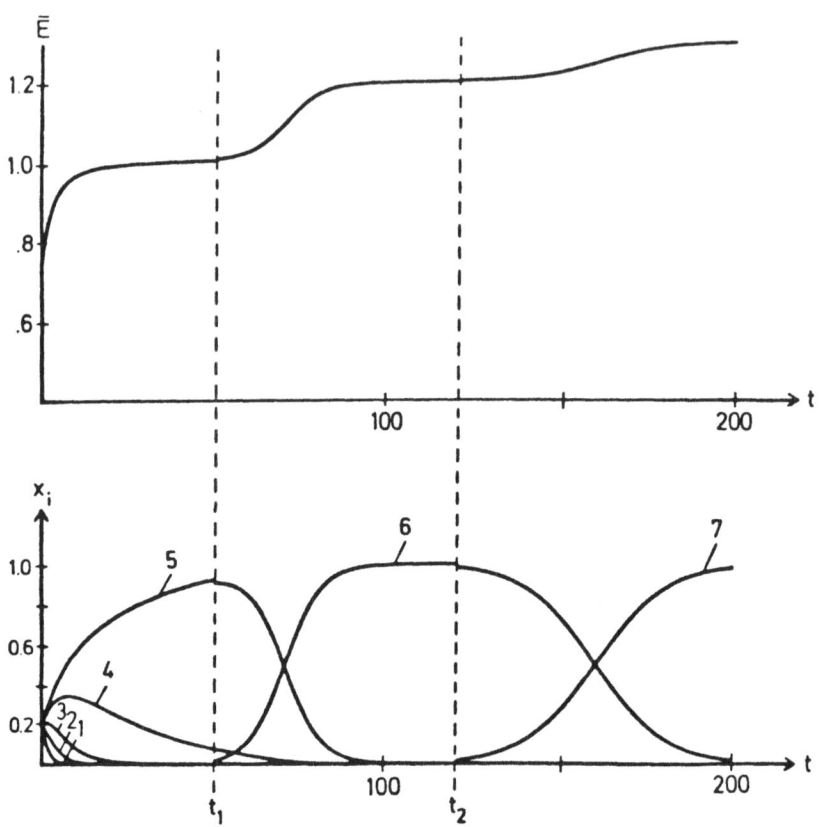

Fig.10. Selection and optimization in an ensemble of polynucleotides
in the evolution reactor. We present solution curves of the differential equation (13) with the n=7 and w_{ij} =0 if i≠j. The values for
w_{ii} were chosen in the sequence 0.3, 0.65, 0.82, 0.96, 1.01, 1.2
and 1.3 for i=1,2,...,7. The initial conditions are :
$x_1(0) = x_2(0) = ... = x_5(0) = 0.2$ and $x_6(0) = x_7(0) = 0$. Thus, the mean excess
production starts from an intial values $\bar{E}$ =0.748 and increases
steadily as the population becomes homogeneous through selection:
$\bar{E} \to 1.01$; $x_1,x_2,x_3,x_4 \to 0$ as $x_5 \to 1$. We recognize easily the case for
the increase of $\bar{E}$: the less efficiently growing polynucleotides
are eliminated. They disappear in the same sequence as their excess
production increases, namely 1,2,..., 5. At t=t_1=50 we observe
a fluctuation $x_6 = 0 \to x_6 = \delta$. The appearance of I_6 as a favourable
mutant leads to a further increase in $\bar{E}$ which approaches the value
$\bar{E}$ =1.2 as I_5 disappears now and the population becomes homogeneous
again : $x_1,x_2,x_3,x_4,x_5 \to 0$ as $x_6 \to 1$. The same story happens when a
more efficient mutant appears at t=t_2=120. I_6 is replaced by I_7
and $\bar{E}$ approaches the temporal optimum $\bar{E}$=1.3. This example illustrates the nature of the selection process and the role of the mean
excess production $\bar{E}$ as the quantity which is optimized.

distribution.Therefore, replication at low accuracy has been called
"random replication" (Swetina and Schuster, 1982). It has another
important aspect which lies outside the deterministic description by
means of (13). The number of possible polynucleotide sequences is
"hyperastronomic" already for polynucleotides of moderate chain
lengths. Thus, the number of possible sequences exceeds by far the
number of individuals in any realizable population and we expect
a severe effect of finite population size. We really cannot have
less than a single copy of a given polynucleotide in the reactor,
but all stationary concentrations will be far less this ultimate
critical limit. Hence, we are dealing with a set of sequences
which changes from generation to generation; new sequences appear
due to copying errors and a certain percentage of the old sequences
disappears as a consequence of degradation and dilution. The notion
of "presence in equal amounts" can be replaced   at best by "equal
probability of realization" in a long term experiment.

It is important to realize that the border between the two
scenarios (1) and (2) is very sharp. There exists a well defined
error threshold below which we have random replication. In figure
11 we show the transition from faithful to random replication
in the case of polynucleotide with ν = 50 bases. For longer
sequences the transition is even sharper. The error threshold relation has been used to explain the lengths of the genome observed
with simple RNA bacteriophages and bacteria. For further details
see Eigen and Schuster (1979).

The third and last example we consider here deals with kinetic degeneracy. In this case we cannot apply the deterministic differential equations of conventional kinetics. Replication is visualized as a stochastic process. Instead of concentrations we consider the probability distribution of particle numbers. The time
dependence of these probability distributions is described by means
of a master equation (see e.g. Mc Quarrie, 1967). Analytical solutions of the master equation corresponding to equation (13) are
not available for the general case. Some examples of numerical
simulations have been published a few years ago (Ebeling and
Feistel, 1977). Another investigation studied expectation values
and dispersions of the probability distribution (Jones and Leung,
1981). Here we shall consider a special case only which is a good
illustration of random selection. We assume conditions which are
the opposite extreme compared to those chosen in the previous
examples : the case of complete kinetic degeneracy, $f_1=f_2=...=f_n=f$
and $d_1=d_2=...=d_n=d$. For technical reasons it is easiest to treat
the case f=d. Mutation terms are neglected ($Q_{kk}=1$, $Q_{kj}=0$ if $j\neq k$).
In addition , we assume the following initial conditions : at t=0
we have n different polynucleotides each one present in a single
copy only. The event at which a polynucleotide sequence becomes

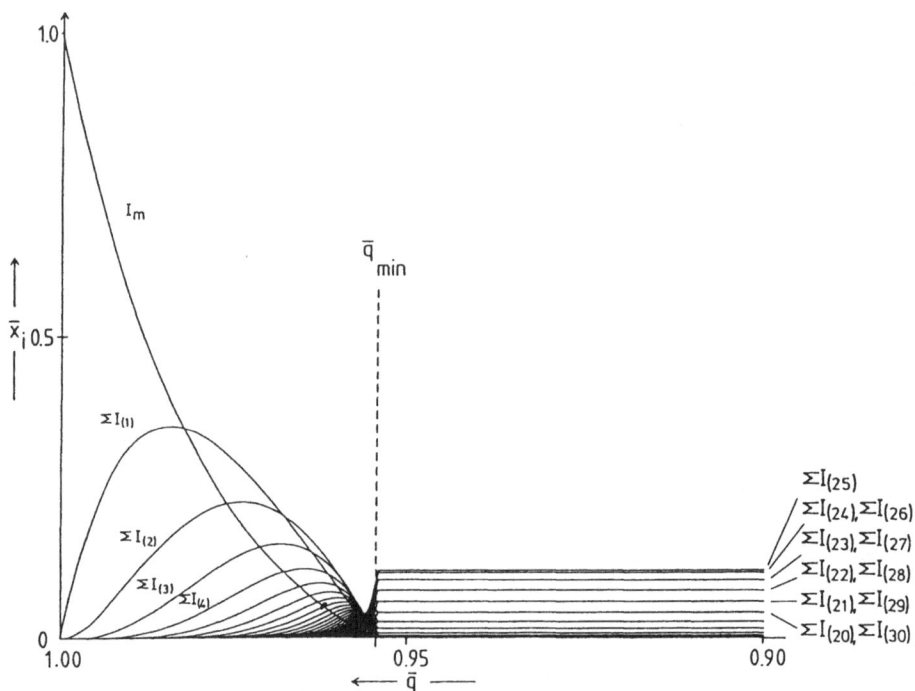

Fig.11. Stationary mutant distribution (quasispecies) in a replication ensemble of polynucleotides. The curves represent the relative concentrations of the master sequence (I_m), the sum of the relative concentrations of all one error mutants ($\Sigma I_{(1)}$), of all two
error mutants ($\Sigma I_{(2)}$)etc., as functions of the mean single digit
replication accuracy ($\bar{q}$). The single digit accuracy is related to
the total accuracy by the relation $Q = \bar{q}^\nu$. In this example the length
of the polynucleotides is $\nu = 50$ bases.The formation rate constant
for the mastersequence is chosen to be $f_m = 10$, for all other sequences $f_1 = f_2 = \ldots = f_n = 1$. All decomposition rate constants are equal ,
$d_m = d_1 = d_2 = \ldots = d_n$, and hence do not enter the differential equation
(5). From these parameter values we calculate $\sigma_m = 10$ and a critical single digit accuracy $\bar{q}_{min} = 0.954$. With decreasing replication
accuracy a pronounced decrease in the relative concentration of
the master sequence is observed. One error mutants, then two error
mutants dominate the polynucleotide distribution. Below the critical
accuracy $\bar{q}_{min}$ the concentrations are exclusively determined by the
statistical weight of the corresponding sequences. Hence, the sum
of the concentrations of 25 error mutants ($\Sigma I_{(25)}$) is largest,
followed by 24- and 26 error mutants ($\Sigma I_{(24)}$, $\Sigma I_{(26)}$) etc. For
further details see Swetina and Schuster (1982).

extinct is considered to be a random variable T_k. The set of random variables T_k, $k=1,2,\ldots,n$, has been called sequential extinction times (Schuster and Sigmund, 1983). The index k denotes the number of different polynucleotide sequences which are present just after the event T_k. Thus, we have n different polynucleotide sequences between the events T_n and T_{n-1}; $n-1$ sequences between T_{n-1} and T_{n-2} etc. In our example the expectation values of the sequential extinction times can be calculated easily :

(17)
$$E\{T_k\} = \frac{n-k}{k} \cdot \frac{1}{f}$$

A numerical example (n=20) is represented in Figure 12. Most of the sequences become extinct within the initial period. Then, the number of different sequences decreases more slowly until we have finally a single sequence which is present in many copies. This process of selection is completely random with respect to the particular sequence which ultimately survives. In case we start from the same numbers of initial copies for the different sequences the a priori chances to survive by random selection are the same for all sequences.

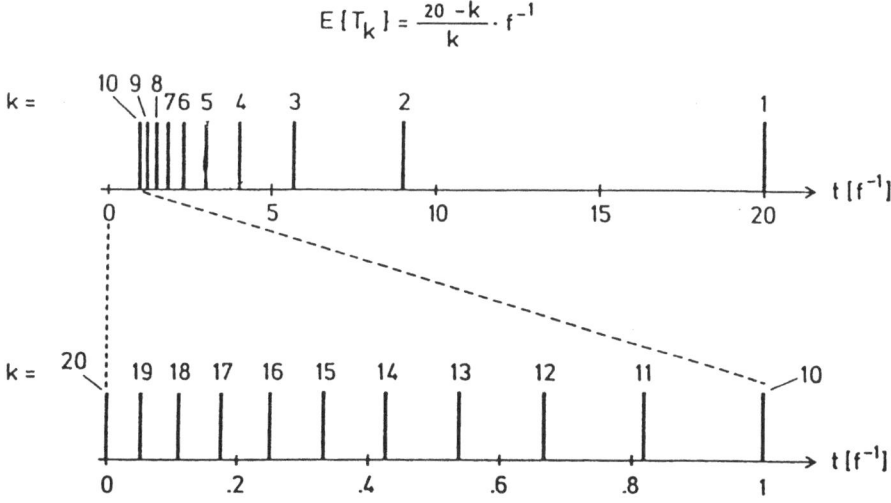

Fig. 12. The distribution of expectation values for the sequential extinction times, $E\{T_k\}$, with n=20.

There is experimental evidence for the occurrence of random selection in polynucleotides replication (Biebricher, 1983) : there are many polynucleotide sequences which are equally well recognized and equally fast synthesized by the RNA polymerase. Hence, all these polynucleotides correspond to optima of the selection process and we are dealing with a case of kinetic degeneracy. Indeed, different de novo experiments lead to polynucleotides which give different fingerprints on enzymatic hydrolysis and gel electrophoresis, thus indicating that different sequences have been selected in different runs.

A purely phenomenological approach, which just counts the number of variants present in the population or just records the concentrations of polynucleotides, in unable to distinguish between adaptive selection and random selection. In order to analyze the nature of the driving force of the selection process we have to go beyond pure phenomenology. In the case of polynucleotides, kinetic studies and sequence analysis show that both selection mechanisms are in operation in serial transfer experiments.

ACKNOWLEDGEMENTS

The work reported here has been supported financially by the "Austrian Fonds zur Förderung der wissenschaftlichen Forschung" projects no. 4506 and 5286. Technical assistance in preparing the manuscript by Mrs.J. Jakubetz and Mr.J. König is gratefully acknowledged.

REFERENCES

Biebricher, C.K. (1973) Evolutionary Biology, 16, 1-52.
Biebricher, C.K., Eigen, M., Luce, R. (1981) J. Mol. Biol. 148,369-390.
Biebricher, C.K., Diekmann, S., Luce, R. (1982)J. Mol. Biol. 154, 629-648.
Biebricher, C.K., Eigen, M., Gardiner, W.C.jr. (1983) Biochemistry 22, 1544-2559.
Ebeling, W., Feistel, R. (1977) Ann. Phys. 34, 81- 90.
Eigen, M. (1971) Naturwissenschaften 58, 465-526.
Eigen, M., Schuster, P. (1979) "The Hypercycle - A Principle of Natural Self-Organization", Springer, Berlin.
Gassner, B., Schuster, P. (1982) Mh.F. Chemie 113 , 237-263.
Hill, D., Blumenthal, T. (1983) Nature 301, 350-352.
Hofbauer, J., Schuster, P. (1984) in "Stochastic Phenomena and Chaotic Behaviour in Complex Systems", P. Schuster ed., Springer Berlin.
Jones, B.L., Leung, H.K. (1981) Bull. Math.Biol. 43, 665-680.
Jones, B.L., Enns, R.H., Ragnekar, S.S. (1976) Bull. Math. Biol. 38, 15-28.

Küppers, B.O. (1979) Naturwissenschaften 66, 228-243.
McQuarrie, D.A. (1967) J. Appl. Prob. 4 , 413- 478.
Schuster, P., Sigmund, K. (1984) Bull. Math. Biol.46, 11-17.
Schuster, P., Sigmund, K. (1984) in "Stochastic Phenomena and Chaotic
Behaviour in Complex Systems", P. Schuster ed., Springer Berlin.
Spiegelman, P. (1971) Quart. Rev. Biophys. 4, 213-253.
Sumper, M., Luce, R. (1975) Proc. Natl. Acad. Sci. USA 72, 162-166.
Swetina, J., Schuster, P. (1982) Biophys. Chem. 16, 329-345.
Thompson, C.J., McBride, J.L. (1974) Math. Bioscience 21, 127-142.

LIST OF CONTRIBUTORS (C) AND CONFERENCE CHAIRMEN (C.C)

T. BÁK (C.C)
Kemisk Laboratorium III
H.C. Ørsted Institutet
Universitatsparken 5
2100 - Kobenhaven Ø
Danmark

A. CORNISH-BOWDEN (C)
Department of Biochemistry
The University of Birmingham
Po Box 363
Birmingham
B 15 2TT - England

K. DALZIEL (C)
Department of Biochemistry
University of Oxford
South Parks Road
Oxford
Ox 1 3QU - England

S. DICKINSON (C.C)
Department of Biochemistry
University of Hull
Hull
HU6 7RX - England

H.B. DUNFORD (C.C)
Department of Chemistry
The University of Alberta
Edmonton - Alberta
T6G 2G2 - Canada

A. GOLDBETER (C)
Université Libre de Bruxelles
Faculté des Sciences
1050 - Bruxelles
Belgique

B. GUTTE (C)
Biochemisches Institut der
Universitat Zurich
Zurichbergstrasse 4
CH-8028 Zurich
Switzerland

B. HESS (C)
Max Planck Institut fur
Ernährungsphysiologie
Rheinlanddam 201
46 Dortmund
West Germany

R.Y. HSU (C.C)
College of Medicine
Department of Biochemistry
State University of New York
Upstate Medical Center
766 Irving Avenue
Syracuse , NY 13210
United States

T. KELETI (C)
Department of Enzymology
Institute of Biochemistry
Hungarian Academy of Sciences
Po Box 7
H-1502 Budapest
Hungary

K. KIRSCHNER (C)
Biozentrum der Universitat Basel
Abteilung Biophysikalische Chemie
Klingelbergstrasse 70
CH-4056 Basel
Switzerland

B. LABOUESSE (C.C)
IBCN-CNRS
1 rue Camille Saint Saens
33077 Bordeaux Cedex
France

B. MANNERVIK (C.C)
University of Stockholm
Arrhenius Laboratory
Department of Biochemistry
S. 106 91 Stockholm
Sweden

K. NEET(C)
Department of Biochemistry
School of Medicine
Case Western Reserve University
2109 Adelbert Road
Cleveland , Ohio 44106
United States

R. PERHAM (C.C)
Department of Biochemistry
University of Cambridge
Tennis Court Road
CB2 1QW Cambridge
England

J. RICARD (C)
CBM - CNRS
BP 71
13402 Marseille Cedex 9
France

G.L. ROSSI (C.C)
Istituto di Biologia Molecolare
Università di Parma
Via del Taglio
43100 Parma
Italy

P. SCHUSTER (C)
Institut fur Theoret. Chemie
und Stranhlenchemie der
Universitat Wien
Wahringerstrasse 17
A. 1090 Wien
Austria

J. STUCKI (C)
Pharmakologisches Institut
der Universitat
Bern
Switzerland

D. THOMAS (C)
Laboratoire de Technologie
Enzymatique
CNRS
BP 233 Compiègne
France

C. VEEGER (C.C)
Agricultural University
de Dreijen 11/6703 BC
Wageningen
The Netherlands

R. WELCH (C)
Department of Biological Scien-
ces
University of New Orleans
Lake Front
New Orleans 70148
United States

E. WHITEHEAD (C)
Istituto di Chemicà Biolo-
gica
EURATOM Res. Group
Università di Roma
Roma 00185
Italy

J.WONG (C)
Department of Biochemistry
University of Toronto
Toronto
Canada

J. YON (C.C)
Laboratoire d'Enzymologie physico-
chimique et Moléculaire
CNRS
91400 Orsay
France